Protein Architecture

The Practical Approach Series

SERIES EDITORS

D. RICKWOOD

Department of Biology, University of Essex
Wivenhoe Park, Colchester, Essex CO4 3SQ, UK

B. D. HAMES

Department of Biochemistry, University of Leeds
Leeds LS2 9JT, UK

Affinity Chromatography
Anaerobic Microbiology
Animal Cell Culture
Animal Virus Pathogenesis
Antibodies I and II
Biochemical Toxicology
Biological Membranes
Biosensors
Carbohydrate Analysis
Cell Growth and Division
Cellular Calcium
Cellular Neurobiology
Centrifugation (2nd Edition)
Clinical Immunology
Computers in Microbiology
Crystallization of Proteins and Nucleic Acids
Cytokines
The Cytoskeleton
Directed Mutagenesis
DNA Cloning I, II, and III
Drosophila
Electron Microscopy in Biology
Electron Microscopy in Molecular Biology
Enzyme Assays
Essential Molecular Biology I and II
Fermentation
Flow Cytometry
Gel Electrophoresis of Nucleic Acids (2nd Edition)
Gel Electrophoresis of Proteins (2nd Edition)
Genome Analysis
HPLC of Small Molecules
HPLC of Macromolecules
Human Cytogenetics
Human Genetic Diseases
Immobilised Cells and Enzymes
Iodinated Density Gradient Media
Light Microscopy in Biology
Liposomes
Lymphocytes
Lymphokines and Interferons

Mammalian Development
Mammalian Cell Biotechnology
Medical Bacteriology
Medical Mycology
Microcomputers in Biochemistry
Microcomputers in Biology
Microcomputers in Physiology
Mitochondria
Molecular Neurobiology
Molecular Plant Pathology I and II
Monitoring Neuronal Activity
Mutagenicity Testing

Neurochemistry
Nucleic Acid and Protein Sequence Analysis
Nucleic Acids Hybridisation
Nucleic Acids Sequencing

Oligonucleotide Synthesis

PCR

Peptide Hormone Action
Peptide Hormone Secretion
Photosynthesis: Energy Transduction
Plant Cell Culture
Plant Molecular Biology
Plasmids

Post-implantation Mammalian Embryos
Prostaglandins and Related Substances
Protein Architecture
Protein Function
Protein Purification Applications
Protein Purification Methods
Protein Sequencing
Protein Structure
Proteolytic Enzymes

Radioisotopes in Biology
Receptor Biochemistry
Receptor—Effector Coupling
Receptor—Ligand Interactions
Ribosomes and Protein Synthesis

Solid Phase Peptide Synthesis
Spectrophotometry and Spectrofluorimetry
Steroid Hormones

Teratocarcinomas and Embryonic Stem Cells
Transcription and Translation

Virology

Yeast

Protein Architecture

A Practical Approach

A. M. LESK

Department of Haematology,
University of Cambridge Clinical School,
Addenbrooke's Hospital, Cambridge

OXFORD UNIVERSITY PRESS
Oxford New York Tokyo

Oxford University Press
Walton Street, Oxford OX2 6DP

Oxford is a trade mark of Oxford University Press

Published in the United States
by Oxford University Press, New York

British Library Cataloguing in Publication Data
Lesk, A. M.
Protein architecture.
1. Proteins. Structure
I. Title
574.19296
ISBN 0-19-963054-2
ISBN 0-19-963055-0 (pbk.)

Library of Congress Cataloging in Publication Data
Lesk, A. M.
Protein architecture: a practical approach/A. M. Lesk.
p. cm.—(The practical approach series)
1. Proteins—Structure. 2. Proteins—Structure—Atlases.
3. Proteins—Structure—Computer simulation. I. Title.
II. Series.
QP 551.L464 1990 547.7'50442—dc20 90-45404
ISBN 0-19-963054-2
ISBN 0-19-963055-0 (pbk.)

Photoset by Cotswold Typesetting Ltd, Gloucester

Printed in Great Britain by
Information Press Ltd, Eynsham, Oxford

Dedicated to Annalisa Pastore and Anna Tramontano

Preface

THERE are many books on protein structure, written from many different points of view. This book is more specialized than most. It does not contain much chemistry; the emphasis is rather on protein structure as three-dimensional pattern. Rutherford said, 'All science is either physics or stamp collecting'; I reply that the investigation of protein structure includes the best qualities of both. Here an effort has been made to show the variety of protein folds, in an intelligible way, and to describe techniques available for drawing pictures useful in analysing protein structures. It is hoped that the pictures contained here will themselves serve as a useful reference, and the reader will be able to make others as needed to investigate or report on particular structures or sets of structures.

For a more complete appreciation of the field of protein structure and function, it is a pleasure to recommend other books by colleagues, notably Schulz and Schirmer (1), Creighton (2), Fersht (3), and Bränden and Tooze (4), of which the last is closest in approach to this one.

Most molecular graphics is done today on workstations with colour screens and equipped with devices for stereo viewing. The inclusion of coloured pictures in this book is necessarily limited, but many of the pictures are printed as stereo pairs. They can be viewed using a standard portable stereo viewer, or, with a little practice in allowing one's eyes to diverge, without one.

The author is grateful to many colleagues who have contributed advice, coordinate sets, or pictures—or in some cases all of the above. During my recent professional activities I have been indebted to the entire community of protein crystallographers who have generated the data with which I work. More particularly I wish to thank Drs C.-I. Bränden, C. Chothia, T. E. Creighton, R. Diamond, (Mr) M. B. Gerstein, T. A. Jones, A. Pastore, O. B. Ptitsyn, A. Tramontano, G. Vriend, and Ms H. Fry and Ms C. Raulfs for help during the preparation of the book.

I thank the Kay Kendall Foundation for generous support.

Arthur M. Lesk

Contents

Contents

5. Structure comparison and structural change

Contents

The plate section falls between pages 106 and 107.

xiv

Overview and background

1 Introduction

Detailed structures at atomic resolution are now known for about a thousand proteins. These results reveal some of the variety of structural patterns that nature produces in this family of molecules. They support a number of interesting and important scientific endeavours, which I sketch here in outline:

1. Interpretations of the mechanisms of function of individual molecules.

The catalytic activity of an enzyme such as the serine protease chymotrypsin can be explained in terms of physical-organic chemistry, on the basis of the interactions of residues of the protein with the substrate atoms around the scissile bond. The size and shape of a pocket in chymotrypsin and in a number of related enzymes explains the variation in their substrate specificities, in this case much according to Emil Fischer's 'lock-and-key' hypothesis. The specificity involves complementarity in shape and electric charge.

2. Approaches to the 'protein folding' problem.

The amino acid sequences of proteins dictate their three-dimensional structures. This is the mechanism by which the one-dimensional genetic code stored in DNA is translated into three dimensions: Nucleotide sequence enciphers amino acid sequence; amino acid sequence encodes three-dimensional conformation. It cannot be said that nature's 'algorithm' relating amino acid sequence to protein structure is yet well-understood; certainly we cannot predict the conformation of a novel protein structure from its amino acid sequence. However, some general principles of protein architecture have become clear, in that the nature of the interactions that stabilize native conformations, and some of their structural implications, have been identified.

3. Patterns of molecular evolution.

There are several families of protein structures for which we know dozens or even hundreds of amino acid sequences, and about 10–20 structures; for example, the globins, the cytochromes c, and the serine proteases. It has been possible to analyse the mechanism of evolution, in that we can observe the structural and functional roles of the sets of residues that are strongly conserved and those that vary relatively freely, and can describe the structural consequences of changes in the amino acid sequence. In general, the response of protein structures to

1

mutations, insertions, and deletions in the amino acid sequence is conformational change; and the greater the divergence of the sequences, the greater the divergence of the structures. However, selection imposes global constraints on the structure to preserve function. It is possible to define a 'core' of the family of structures that tends to be well-conserved during evolution. When proteins evolve with changes in function, these constraints on the structure are relaxed—or rather, replaced by alternative constraints—and the sequences and structures can change more radically.

4. Prediction of the structures of closely-related proteins.

Observed relationships between the divergence of amino acid sequence and divergence of protein structure in families of homologous proteins makes possible the prediction of protein structure from amino acid sequence—in favourable cases and to a limited extent. (The caution and qualification are *not* exaggerated!) Suppose someone determines the amino acid sequence of a new protein, and the sequence shows it to be related to one or more proteins for which we know both the sequence and the structure. As a rule of thumb, if no more than 50–60% of the residues have changed between the unknown, target protein, and its nearest relative of known structure, then the nearest relative of known structure will provide a reasonable quantitative model for almost all of the unknown protein.

5. Protein engineering.

Protein biochemists used to be like astronomers, in that they could observe but not alter the subjects of their studies. Now, with techniques of genetic engineering, it is possible to design and test modifications of known proteins (and perhaps some day even to design novel ones). Potential applications include:

(a) Modifications to probe mechanisms of function, such as the changes in residues involved in enzyme–substrate interactions in *Bacillus stearothermophilus* tyrosyl-tRNA synthetase, or the construction of 'Mermaid haemoglobin': a hybrid of mammalian and fish chains to test theories of the Bohr effect (the dependence of oxygen affinity on pH).

(b) Attempts to enhance thermostability; for example, by introducing disulphide or salt bridges. Thermostable enzymes would have industrial application as ingredients in laundry products to permit higher wash temperatures.

(c) Clinical applications, such as the transfer of the antigen-binding loops from a rat antibody to a human antibody framework, to produce a molecule that retains a known therapeutic activity in humans but reduces the side-effects arising from the patient's immunological response against the rat protein.

(d) Modifications to investigate the roles of different residues in stabilizing the native structure or intermediates in folding. These include 'saturation mutagenesis', to explore the repertoire of sequences consistent with a natural structure, and the characterization of folding intermediates by the effects of mutation on folding kinetics.

2

(e) Modifying antibodies to give them catalytic activity. Two features of all enzymes are: ability to bind substrate specifically, and the juxtaposition of bound substrate with residues that catalyse a chemical change (by stabilizing the transition state of a reaction). Immunoglobulins provide the binding and discrimination; the challenge to the chemist is to introduce the catalytic function.

6. Drug design.

There are many proteins specific to pathogens that we want to inactivate. Knowing the structure of the AIDS protease, or the coat protein of influenza virus, it should be possible to design molecules that will bind tightly and specifically to an essential site on these molecules, and interfere with their function.

The scientists who participate in these studies come from a variety of fields: biochemists, molecular biologists, crystallographers, computer specialists. Each has his or her own set of tools: incubators, columns, gel readers, diffractometers. But one facility they all share: if you wait in the right place, you will see them emerging from a small, dark room, slightly unsteady on their feet, blinking or even rubbing their eyes, after a session with a special computer device, equipped with an appropriate program system, that has permitted them TO LOOK AT THE STRUCTURES.

2 Molecular graphics

Topics from the field of molecular graphics form the subject of this book. Full coverage of the subject would include a description of the facilities available (both 'hardware' and 'software') and a description of how to use them. In fact, there is quite a large literature on computer graphics, but relatively little aimed specifically at training molecular biologists to use molecular graphics systems effectively. Manuals describe how to use programs, but not what to do with them. It is not so easy to decide what to look at; or how to design an intelligible illustration tailored to the scientific question one wants to address. I have concluded that I can best serve the reader by emphasizing the 'molecular' rather than the 'graphic' aspects of the field; however, this section contains a more general summary of the subject.

It will emerge also that because the state of the technical art is changing so fast, it is appropriate to emphasize general principles which will survive, and to recognize and concede that descriptions of specific systems will not. Still another disclaimer: this discussion is aimed primarily at people who wish to *use* molecular graphics systems to investigate questions of protein or nucleic acid structure. Those who wish to create their own graphics system should know what is here, but will require additional technical information readily available in the specialist literature.

For texts dealing with the general topic of computer graphics, see, for example, refs 5–10.

2.1 Graphics devices: 'hardware'

A graphics device is an instrument for creating a picture by a computer. Devices can be classified according to the nature of the pictures that they can produce, and the speed with which they can produce them. We shall see, for example, that only a device that can simulate adequately quickly the rotation of a molecule can create the illusion that one has an object in one's hand which one can turn continuously, to explore different orientations. This 'cinematographic threshold' is about 15 frames per second. (Commercial cinema runs at 24 frames per second.)

2.1.1 Vector vs. raster devices

One fundamental distinction is that between *line drawings* (or vector graphics or calligraphic graphics) and television-like images, or *raster graphics*. In chemical terms, if we represent a molecule by showing each chemical bond as a line segment we are doing calligraphic graphics (*Figure 1.1*), whereas if we represent each atom by a shaded sphere we are doing raster graphics (*Figure 1.2*).

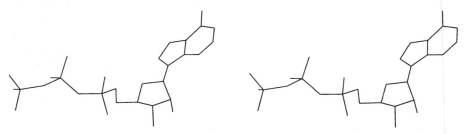

Figure 1.1 Adenosine triphosphate (ATP). A line drawing represents a molecule by a collection of line segments corresponding to the chemical bonds. Bonds to hydrogen atoms are omitted.

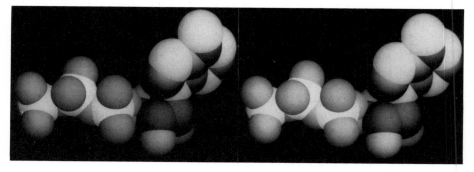

Figure 1.2 A space-filling representation of ATP, in the same orientation as *Figure 1.1*. Each atom is represented by a sphere. Hydrogen atoms are omitted.

Once upon a time, the technology of vector graphics and raster graphics were fairly well separated: For vector graphics one had pen plotters—now usually superseded by laserprinters—or specialized devices with electronic screens; the best known of these were Tektronix terminals and the Evans & Sutherland Picture System (PS-300) series.* For raster graphics, there were many devices that would map images on to a screen: typically each point (or pixel) in a 1000×1000 (approximately) array was specified by one byte (= eight bits = an integer in the range 0–255), and a program could translate any of the 256 possible values into a specified intensity and colour on the screen. For example, one could display any point in any of 8 different colours, with 32 intensity levels for each colour. (In truly ancient times one could produce raster images on a line printer by overprinting selected sets of characters to create different densities; for an application to molecular graphics see ref. 11.) It is unnatural and difficult, although not entirely impossible, to create raster images on a vector device. Conversely, typical raster devices can draw lines, but fine detail is impossible and oblique lines are inevitably jagged. Evans & Sutherland have developed a special raster device adapted to drawing lines of very high quality; this technology, called Shadowfax, is used in their PS-390 series and newer workstations.*

This split between vector and raster hardware was reflected in a disparity in the computer software, and especially in the associated data structures. 'Computer graphics' was the field of hardware and algorithms for vector graphics, and 'image processing' was the field of hardware and algorithms for raster graphics.

For vector graphics, plotting devices produced images on paper, but not interactively. Slow line-drawing screens such as the Tektronix 4014 could not draw pictures fast enough to simulate real-time rotation. Specialized hardware, of which the Evans & Sutherland PS-300 series has been the most common example, could perform real-time rotation and interactive manipulation of line drawings.

For raster graphics, the most common devices are able to display a 1000×1000 point image (approximately), in which each point is selected from one of 256 colour-intensity values. Given a reasonably high-speed link between the screen and the host computer, the time required to create and display an image was limited by the execution time of the program in the host, typically 1–2 min, depending of course on the speed of the host and the complexity of the picture. But even the display of a pre-calculated picture required at least a few seconds, precluding real-time rotation. Newer devices, incorporating special hardware, have reduced the time to create a new raster image to a few seconds (in the case of the Evans & Sutherland PS-390) and, in the newest devices, to below the cinematographic limit.

Although molecular graphics has always made use of both vector and raster systems, for many years the most advanced work was done on vector systems, because only for these was it possible to simulate real-time rotation, that is, only

---

* PS 300, PS 390, and Shadowfax are trademarks of Evans & Sutherland Computer Corporation.

these allowed the observer to rotate an image as fast as a knob could be turned, giving the illusion of manipulating an object held in the hand.* Standard program packages such as FRODO (by T. A. Jones) are installed in many institutions on the Evans & Sutherland Picture System 300 series. However, the newest generation of graphics devices is capable of producing *both* vector and raster images on the same screen, and the prices have come down to the point where an individual laboratory can afford them. The most powerful of such devices are able to rotate a shaded-sphere image of an entire protein in 'real-time'. Thus, real time rotation of raster images has now entered molecular graphics laboratories. Moreover, the facility for mixing vector and raster graphics on the same screen creates possibilities for novel combinations of representations. The challenge to the designers of software is to take advantage of these facilities in a genuinely scientifically useful way.

2.1.2 Colour

The fundamental problem of molecular graphics is to make complex structures intelligible. Colour is an important tool, because it permits similar-looking items within a complex picture to be distinguished visually. For example, in a line-drawing of the haemoglobin molecule, it might be useful to draw the α-chains in red and the β-chains in green. In a raster picture showing a shaded-sphere representation of a protein, it is useful to distinguish different types of atoms by different colours, possibly (but, of course, not necessarily) according to the assignments used in the familiar Corey–Pauling–Koltun (CPK) models. Beyond this, it is often useful to show selected residues, or ligands, in a distinguishing colour.

One of the most important advantages of molecular graphics over physical models is the ability to superpose or overlay molecules in one picture, so that the structural differences can be recognized and identified. Showing the different molecules in different colours is the best way to keep different molecules distinct.

Intelligent use of colour is an important element in the design of intelligible pictures, and this subject will recur frequently in our discussion.

2.1.3 'Hard copy': harvesting the pictures

Interactive sessions at a fancy screen are essential, but it is also important to preserve and record the images produced. There are several reasons for this:

1. Work at a graphics terminal often does not lend itself to reflection: someone else has booked the next hour and one is under pressure to finish on time; and the rooms are often dark, cold, noisy, and uncomfortable. Despite the power of the equipment, most sessions at graphics terminals are 'nasty, brutish, and

---

* Raster devices that simulated real-time rotation did exist during this period, but were so expensive that their applications were limited to military applications, or to the training of aircraft or supertanker pilots. The potential of such devices for molecular biology was not unrecognized, of course. (The author once gave a talk entitled 'The line will decline when the raster gets faster'.)

short'. Just as with restaurants, people find it important to be able to 'take-away' pictures. These are studied at leisure, in conjunction with notes and with other pictures from other sessions. (Possibly the new generation, brought up on multi-hour sessions with computer games, will not suffer from this problem.)

2. The results of the investigation must be preserved. Often the successful result of a graphics session is a single picture that contains exactly the right representations of the right sets of residues, to reveal what one has discovered.

3. It is often desired to published illustrations of structures.

Here are some of the possibilities:

When graphics was done exclusively with a plotter, all 'copy' was 'hard'. Multi-pen plotters can produce different colours, although their palette is limited compared to modern colour screens, and of course they can only produce line-drawings. Laser printers have until recently been limited to black-and-white drawings, although colour laser printers are now available but not yet common. Often the data structure associated with an interactive graphics screen is different from that of the plotter or laserprinter so that some software is required to make the conversion.

It is obvious that the most direct way of capturing the image from a screen (vector or raster) is to photograph it. One can mount a camera in front of the screen and shoot away; this requires some calibration of the exposure, and requires the co-operation of other people in the room (even more problematical are those who wish to enter and leave it). Such problems can be avoided by a *camera station*, which is a closed box containing a camera, a screen 'slaved' to the display, and a light meter; the exposure can be under computer control.

Movies are difficult without special equipment. If a movie camera is simply placed in front of the screen of a device producing an animation sequence (for example, a series of pictures showing a molecule rotating about an axis) one will encounter the problem of the frequency difference between the refresh rate of the screen and the shutter frequency of the camera. It is essential to have a movie camera that can be run in single-frame mode, under computer control. Movie cameras can be mounted on camera stations. Many modern graphics devices have output ports for video signals which can be recorded on tape. Many of us are waiting impatiently for High Definition TV.

2.2 Depth perception

All graphics devices except holograms portray three-dimensional structures as two-dimensional images, whether on paper or on a screen. It is essential to recapture the lost dimension.

Several techniques are available to help the viewer grasp spatial relationships in the direction perpendicular to the image plane.

2.2.1 Stereoscopic images

The idea behind stereo is that one's right and left eyes view an object from slightly different positions; this is an important component of depth perception in everyday life. The tasks of the graphics system are (a) to prepare two separate images—a left-eye view and a right-eye view—and (b) to deliver only one of these images to each eye.

To prepare the two images, it has been found to be sufficient to rotate the standard image by approximately ± 3 degrees around a vertical axis, assuming an upright observer. (This is not a geometrically exact simulation of one's visual field but it is close enough to work. For discussion see Johnson (12).) Why 3 degrees? The distance between the eyes of an adult is typically about 6.5 cm, and taking a convenient viewing distance from the eyes to an object as 60 cm,

$$\arctan \left(\tfrac{1}{2} \times 6.5 \text{ cm}/60 \text{ cm}\right) = 3.1 \text{ degrees.}$$

There are several ways to deliver only the proper image to the proper eye.

1. The images may be printed side by side, about 6.5 cm apart, and viewed with a standard desk-top stereo viewer. Special viewers are available to view pairs of 35 mm slides containing left and right images. Use of stereo pairs of slides in lectures requires special projection equipment. The author has never had any success using it—at best it creates a distraction while polarized glasses are distributed and the equipment is adjusted (distractions from which the attention of the audience is unlikely ever to be recovered). However, I have heard that there are people who have had better results.

2. The images may be shown side-by-side, at any distance, and viewed through an appropriate arrangement of periscopes to bring the separate images about 6 cm apart. This is often done on the left and right halves of a graphics screen.

3. Instead of displacing the images in space, they can be displaced in *time*. Given a fast enough graphics screen, it is possible to display left- and right-eye images alternately. To view the images, a user may wear a pair of glasses containing electronic shutters synchronized with the display. Alternatively, a screen that changes its direction of polarization in synch. with the alternation of images may be mounted in front of the display, and the viewers wear glasses that have polarizers mounted at different angles before each eye.

Stereo can be combined with perspective projection (see Sections 2.2 and 2.2.1 in chapter 2).

2.2.2 Hidden line and hidden surface removal

In ordinary life, most objects are opaque and that is how we know that one is in front of another. In *Figure 1.2*, only the front surfaces of the frontmost atoms are visible—the picture simulates opaque atoms. Similarly, in a line drawing containing depictions of objects that might logically be imagined as solid, it is possible to simulate their opacity by deleting from the picture lines that pass behind them: This is called 'hidden-line removal'. As an alternative, it is possible to suggest *translucency* by converting lines that pass behind 'solid' objects from solid to broken lines. *Figure 1.3* shows stereo pictures of sperm whale myoglobin, its helices depicted as cylinders, with the cylinders treated as (a) transparent, (b) translucent, and (c) opaque.

Unfortunately, hidden-line conversion is a fairly time-consuming process, and cannot be accomplished in real time on calligraphic devices. Real-time hidden surface removal in raster pictures is possible in the new generation of devices.

2.2.3 The kinetic depth effect and depth cueing

Special hardware in interactive graphics devices permits two additional aids to depth perception. The first is to simulate the appearance of a molecule in a state of rotation. From the comparison of the temporal succession of images, the visual system can extract cues about the spatial relationships in the object. This is called the kinetic depth effect. It is a common frustration of users of interactive graphics devices that as one is searching for the best orientation of a molecule by turning it about under manual control, the kinetic depth effect clarifies the spatial relationships; but when a desired orientation is found and one stops turning the molecule, the image loses its sense of three-dimensionality and the view is not at all as perspicuous as when one was moving through it.

Depth cueing is a modification of the image to cause parts of the depicted structure to fall off in intensity as their distance from the viewpoint increases.

Modern interactive graphics devices permit the simultaneous use of stereo, real-time rotation (the kinetic depth effect) and depth cueing.

2.3 Interaction with the picture

An interactive computer graphics session may be thought of as an activity with two participants: the computer and the molecular biologist. Each is doing things on his or her or its own: the scientist is thinking about what the session has revealed and how to explore the molecules further, and the computer is pushing coordinates through transformations as fast as it can, to support the dynamic image. In addition to these independent activities, and the requirement that both participants be adept at their individual activities, the effectiveness of the session depends on the capacities of the channels by which the two parties communicate with each other. The computer communicates with the scientist primarily through the display, by changing the image. For the scientist to communicate adequately quickly to the computer, a variety of input devices must be provided. These may include the following (in order of the number of dimensions in the data that each device transmits):

1. A set of buttons or function keys (1 bit each).

2. A set of dials (1 scalar each).

9

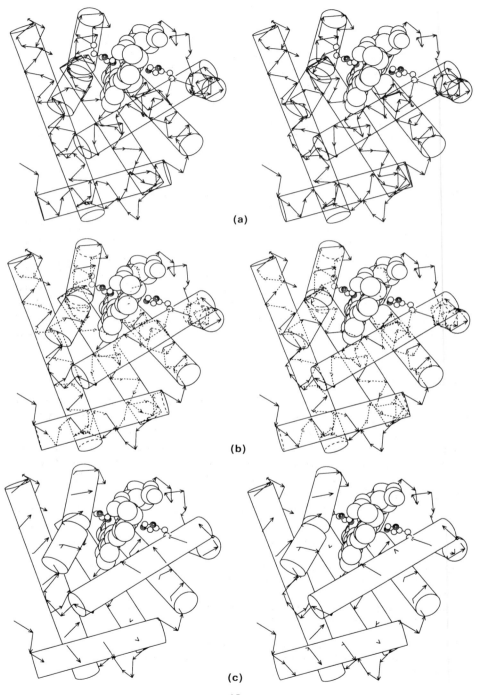

(a)

(b)

(c)

3. A tablet, or light pen, or touch-sensitive screen (a two-component vector, i.e. (x, y)).

[Note: a 'mouse' is a tablet plus a small number of buttons (typically 1–4). A tablet or mouse is often used in connection with a 'cursor' on the screen that follows and reports its position; for example, the mouse and cursor can be used to select options from a 'menu' on the screen.]

4. A 'joystick' [a two- or three-component vector, (x, y) or (x, y, z)].

5. A 'space ball'* [a six-component vector, which could be interpreted, for example, as (x, y, z, R_x, R_y, R_z) where x could be a translation and R_x a rotation].

6. A keyboard, for typing commands.

It is up to the software to interpret the signals from these interactive input devices in terms of effects on the display. Thus, many people find it natural to use the motion of a pen across a tablet to produce a translation of the image on the screen, and to use the rotation of a knob to control the rotation of an image around an axis. But the input signals can be interpreted in any way that the programmer wishes to provide. Obviously, the more interactive input devices the merrier. A typical configuration might include 36 buttons, 8 knobs, a tablet, and a keyboard. One can provide several possible assignments of knobs, using a button to switch among them. ('If only we had 16 knobs instead of 8!', complain the programmers and the users. 'If we gave you 16, you'd want 32!', reply the manufacturers and the budget controllers.)

2.4 Summary

With this we conclude our survey of hardware facilities for molecular graphics. It is worth emphasizing again that the field is in a state of very rapid flux, with many new devices appearing. Both the most expensive and the less expensive devices are increasing dramatically in power. Any fixed sum of money buys much more powerful graphics facilities this year than it did last year, and the trend is likely to continue. Of course, this has the advantage that more graphics facilities are becoming available to more scientists.

As we turn to software and applications, be aware that the difficulties in

---

* Spaceball is the registered trademark of Spatial Systems Pty. Ltd, Sydney, Australia.

Figure 1.3 Sperm whale myoglobin [1 MBD], a schematic diagram in which helices are shown as cylinders, illustrating the effect of hidden-line removal.
(a) No hidden line removal—cylinders transparent
(b) Conversions of hidden lines to dashed lines—cylinders 'translucent'
(c) Removal of hidden lines—cylinders opaque

Note: Protein structures are identified wherever possible by the Protein Data Bank mnemonic in square brackets. Appendix I contains references to publications describing the structure determinations.

describing the state of the art of graphics devices, at a time when many new products are being announced, also create problems for program designers: For which system should we create software? What facilities should we anticipate? Should we make programs very specialized, to take advantage of particular features of some machine, or should we make less than optimal use of the resources of one machine, in order that the programs be more easily portable when the next machine comes along?

Fortunately, most users do not need to worry about such questions.

3 Molecular graphics program systems

Because all succeeding chapters of this book will be concerned with graphics software, it is necessary to describe only some general principles here.

A molecular graphics program can be thought of as sitting within a triangle. At one corner is the computer equipment: a host computer plus specialized graphics devices. At a second corner is the data: the coordinates of the solved protein structures and any information that can be derived from them; or the experimental data from which it is hoped to solve a new protein structure. At the third corner is the scientist.

The scientist wants to use the graphics device to explore the data, and the job of the program is to provide the scientist with effective and versatile facilities to extract information from the database, and create displays. (Ideally the graphics software should be embedded in a more general information retrieval system, but we will return to this point later.)

There are several steps in creating a useful display.

1. Extraction of the proper sets of residues from the database. (An entire protein? Selected regions from a single protein? Selected regions from several proteins?)

2. Choice of a representation. (How much detail: Cα only? All backbone atoms? Backbone and side-chains? Schematic representations? Assignments of colours?) Obviously the choice is constrained by the facilities of the graphics device being used, but the software should allow the scientist to take advantage of what is available. We shall see that it is important to be able to 'fine-tune' the representations of different parts of a molecule. For example, in *Figure 1.3* the haem group and the proximal and distal histidines are shown in greater detail than the rest of the molecule.

3. Once an initial picture is created, to give the scientist as much control as possible to manipulate or modify or interrogate it.

It follows from this that all graphics programs must be able to read selected coordinates, read commands defining representations, translate the coordinates to a picture according to the representations specified, display the picture, and continue to accept commands.

Molecular graphics programs differ in (a) the nature of the working environment they create for the scientist, (b) the versatility of the representations they can create, and (c) the kinds of auxiliary calculations they permit the scientist to perform. For example, the popular programs FRODO (T. A. Jones) and HYDRA (R. Hubbard) both permit the user to display line drawings of the backbones and side-chains of two or more proteins. Both permit interactive rotation of the molecules displayed, interactive changes in colour, and identification of an atom by 'touching' it with the cursor which causes a label to appear. FRODO but not HYDRA provides facilities for the manipulation of the molecule to fit an experimental electron density map. HYDRA but not FRODO permits the computational least-squares superposition of selected regions of the molecules displayed.

Just as in the case of hardware, software is in a state of flux also. People have written many packages, for many different devices, and many new ones are being written. Some of these packages are readily available, some are available but very costly, some are proprietary (see Chapter 6). This book will concentrate on facilities and how to apply them. The reader who puts down this book and enters that dark room will have to study the manuals of whatever system his or her institution provides, but the hope is, that having made the general concepts clear and familiar, it will be an easy transition from a grasp of the principles to a grasp of the joystick.

4 The known protein structures

4.1 Methods of structure determination

The protein structures we know now have almost all been determined by X-ray crystallography. A few contain details derived from neutron diffraction and a few have been determined from nuclear magnetic resonance. One, tobacco mosaic virus, has been determined by X-ray fibre diffraction, and another, bacterio-rhodopsin, has been determined by electron diffraction. It is not appropriate here to describe experimental methods of structure determination in detail, but we may briefly allude to those aspects that affect our discussion. Molecular graphics plays a direct role in X-ray structure determination, in fitting a model to the measured electron density. And in order to analyse the known structures intelligently, it is important to understand the significance of the resolution and refinement of a structural result.

The X-ray structure determination of a protein begins with the isolation, purification and crystallization of a new protein. If a suitable crystal is placed in an X-ray beam, diffraction will be observed, arising from the regular microscopic arrangement of the molecules in the crystal (13). By measuring the intensities of the diffracted rays, and supplementing this set of intensities with additional measured data (classically, additional sets of intensities from chemically modified crystals; and, more recently, measurements of the wavelength dependence of the diffraction intensities) it is possible to create a three-dimensional electron density map of the crystal. Ideally, this contains distinct peaks at the positions of atoms, and a complete atomic model can be built into the appropriate peaks. The computer program FRODO that has been developed to facilitate this process will be described in a later chapter.

Once a complete model is available, a process of refinement ensues, in which by a least-squares fitting procedure the atoms are shifted about to maximize the agreement between the model and the experimental data. In addition to the atomic coordinates, a thermal parameter, or '*B*-factor' is usually included, to indicate the relative mobility of an atom: in effect it measures the size of the atomic peak in the electron-density. In X-ray crystallography, an appropriate model for interpreting the data is that each atom is a collection of electrons occupying a volume of finite size. For small molecules, which form very well ordered crystals, the effective size of the atoms arises primarily from thermal vibrations, and it is possible to measure their amplitudes; indeed, by collecting data at low temperature—by 'freezing out the vibrations'—it is even possible to determine changes in atomic electron density distributions arising from the formation of chemical bonds. For proteins, however, it is more appropriate to regard *B*-factors merely as empirical parameters as they include the effects of disorder as well as vibration, and the reader is cautioned to think twice before venturing to enquire too deeply into the multitude and variety of sins that they may in the most unfavourable cases cover.

Cycles of least-squares refinement may be alternated with sessions at an interactive graphics device to rebuild manually some or all of the model. In many cases, a combination of refinement with molecular dynamics (the simulation of the motion of the molecule according to the rules of classical mechanics) can speed up this process substantially. In the molecular-dynamics-refinement procedure XPLOR (14), the sum of the absolute differences between diffraction intensities and those computed from the current model serves as a pseudo-potential restraining the motion of the particles.

The quality of the structural information derivable from X-ray crystallography is not as good for proteins as it is typically for small molecules, in which the ratio of measured observables to parameters to be determined is so much higher.

Nevertheless, the technique of protein crystallography has advanced greatly in recent years, through the development of area detectors for data collection, synchrotron sources for higher beam intensity, more powerful computer-graphic tools for fitting models, and the application of molecular dynamics to refinement. In many cases, preparation of suitable crystals is the rate-limiting step (15, 16).

In protein structure determination by nuclear magnetic resonance, the experimental data provide a set of distance constraints: typically, a list of protons that are close together in space. This specifies the secondary structure, and gives indications of tertiary interactions. A number of computational approaches, including molecular dynamics, now exist for determining structures consistent with the observed distance constraints.

In either case, what comes out of the structure determination is a more or less complete set of atomic coordinates for the protein. In X-ray crystallography, in unfavourable cases some of the molecule is disordered, invisible in the electron density map, and must be omitted from the model. In NMR structure determination, in unfavourable cases there may be insufficient distance measurements linking parts of the structure, so that long-range spatial relationships may be uncertain.

4.2 How accurate are the structures? Resolution, *R*-factor and stereochemical criteria

It is unfortunate that it is not possible to state with certainty the limits of probable error in the atomic coordinates produced by X-ray crystallography. There are nevertheless a number of fairly reliable indicators. Some of these are derived in the process of the structure determination and reflect experimental observations; others are inherent in the atomic coordinates.

The *resolution* of a structure determination is a measure of how much data was collected. The more data collected, the more detailed the features in the electron density map to be fitted, and, of course, the greater the ratio of observations to parameters to be determined (that is, the atomic coordinates and *B*-factors). Resolution is expressed in ångstroms, and the *lower* the number the *higher* the resolution. Protein structures are generally determined to a resolution between 1.7 and 3.5 Å; those determined at 2.0 Å or better are considered high-resolution.

Confident determination of different types of structural features is dependent on different thresholds of resolution (*Table 1.1*). It should be emphasized that the entries in *Table 1.1* are very rough estimates and depend very strongly on the *quality* of the data.

The *R*-factor of a structure determination is a measure of how well the model reproduces the experimental intensity data. Other things being equal, the lower the *R*-factor the better the structure. The *R*-factor is a fraction expressed as a percentage; $R = 0\%$ would be an impossible ideal case—no disorder, no experimental error; $R = 58\%$ for a collection of atoms placed randomly in the

Table 1.1. Confidence in structural features of proteins determined by X-ray crystallography

Structural feature	Resolution			
	5 Å	3 Å	2.5 Å	2.0 Å
Chain tracing	–	Fair	Good	Good
Secondary structure	Helices fair	Fair	Good	Good
Side-chain conformations	–	–	Fair	Good
Orientation of peptide planes	–	–	Fair	Good

unit cell of the crystal. A typical, well-refined protein structure, based on 2.0 Å resolution data, might have an *R*-factor of less than 20%.

Now that one has seen enough well-determined protein structures to know what they should look like, it is possible to subject atomic coordinate sets to a scrutiny independent of the experimental data. Good protein structures show the following: (1) They are compact, as measured by their accessible surface area and packing density; that is, the buried atoms fit together well, with relatively little empty space between them; (2) they have hydrogen bonds of reasonable geometry, with few hydrogen bonds 'missing' in places where they would be expected, e.g. secondary structures; (3) the distribution of backbone conformational angles in a Sasisekharan–Ramachandran plot is almost entirely confined to the allowed regions (*Figure 1.4*).

4.3 The protein data bank

The Protein Data Bank (PDB) is the collection of publicly available structures of proteins, nucleic acids, and other biological macromolecules (17).

A small group of dedicated individuals at Brookhaven National Laboratory in New York State, USA, under the direction of Dr T. F. Koetzle, has for years carried out the archiving and distribution of the data. Satellite centres in Japan and Australia redistribute the data in their geographical areas; recently data have become available over computer networks.

Currently (July 1990), the PDB contains 503 sets of coordinates, which can be classified into: (1) protein structures determined by X-ray or neutron diffraction or NMR, which may include co-factors, substrates, inhibitors or other ligands; (2) oligonucleotide or nucleic acid structures determined by X-ray crystallography; (3) carbohydrate structures determined by X-ray fibre diffraction (note that a few of the protein structures also contain oligosaccharide); (4) hypothetical models of protein structures; and (5) bibliographic entries.

There is a certain amount of duplication among the entries: In some cases there has been a redetermination of a structure, in different cyrstallization conditions or at higher resolution. In some cases the structure of a molecule has been determined in different states of ligation. In some cases the structures of very closely related molecules have been determined. This makes it difficult to state precisely how many unique protein structures the data bank contains but this number may be estimated at 300. Appendix I contains a catalogue of the structures in the July 1990 distribution.

Each set of coordinates deposited with the Protein Data Bank becomes a separate entry, and is assigned an identifier, for example 1HHO for Human Oxyhaemoglobin. The first character is a version number. Thus, if the crystallographers who deposited the data set 1HHO in the PDB redetermined this structure at higher resolution, they might provide another data set 2HHO. An identifier beginning with the numeral 0 signifies that the entry is purely bibliographic and contains no coordinates.

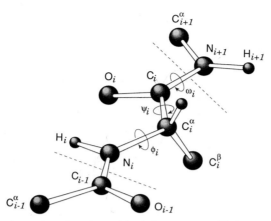

Figure 1.4 The conformation of a polypeptide chain can be specified to a rough approximation in terms of the angles of internal rotation within the backbone. In (a) these angles are defined: the angle ϕ is a rotation around the N–Cα bond, ψ is a rotation around the Cα–C bond, and ω is a rotation around the peptide bond itself. (Usually $\omega \cong 180$ degrees, the *trans* conformation. This figure shows a fragment of polypeptide in the fully extended conformation, with $\phi = \psi = \omega = 180$ degrees.)

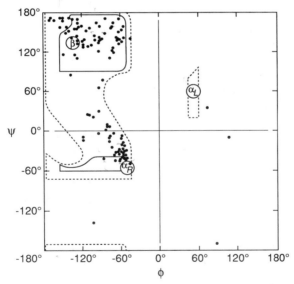

(b) The regions of sterically 'allowed' values of ϕ and ψ, assuming $\omega = 180$ degrees. Solid lines enclose regions estimated to include the maximum tolerable limits of steric strain. The diagram indicates the conformations of three recurrent conformational patterns: the right-handed α-helix, α_R, the β-sheet, β and the left-handed α-helix, α_L. (Residues in the α_L conformation are primarily glycines, with a few asparagines.) Such charts were first developed by V. Sasisekharan and G. N. Ramachandran. On this diagram are plotted the distribution of conformational angles for the residues in a protein structure solved at high resolution and well-refined: ribonuclease A(7RSA). Note that most of the points lie within the allowed regions.

The coordinates are subjected to a set of standard stereochemical checks, and translated into a standard entry format. This format includes information about the structure determination (e.g. the unit cell dimensions and symmetry, the resolution, the R-factor) bibliographic references to papers describing the structure, and the atomic coordinates. Most programs created for molecular graphics or other computational analysis of protein structures can read files in Protein Data Bank format. (See also refs. 18 and 19.)

The Protein Data Bank contains purely bibliographic entries corresponding to structures described in articles but not deposited, structure factors for some proteins (these are the experimental X-ray diffraction measurements from which the structures were determined), and a few computer programs. A separate database created by G. Gilliland at the PDB stores conditions of protein crystallization.

The *Protein Data Bank Newsletter* is currently published four times a year. It contains a catalogue of current holdings, ordering information, and short news items; no one with a serious interest in the field should be without it. For a list of other available documents, including the description of the file formats, see ref. 17.

Line drawings: skeletal and ball-and-stick models

1 Introduction

Traditionally, drawings of small molecules have either represented each atom by a sphere, or each bond by a line segment. Bond representations give a clearer picture of the topology or connectivity of a structure: Kekulé's famous dream in which he saw benzene as a ring of six snakes is a case in point. As we shall see later, what bond representations do not do is to show the space-filling or packing attributes of the atoms.

In this chapter we shall discuss drawings composed of line segments, which include both the pure '1 bond = 1 line segment' skeletal models, and the generalization to 'ball-and-stick' representations, in which both bonds and atoms are shown. *Figures 2.1* and *2.2* contain examples showing the α-helix and β-sheet. Provided hidden lines are removed, by varying the radii of the atoms from zero to the actual Van der Waals radius one can produce a continuum of pictures in which at one extreme only bonds appear and at the other extreme only atoms appear.

Drawings constructed of line segments were the earliest examples of computer-generated molecular graphics, when plotters were the only reasonable output device. Of the many programs produced in this era, by far the most sophisticated was the Oak Ridge Thermal Ellipsoid Program (ORTEP) written by C. K. Johnson (12, 20). This program had many options, including skeletal and ball-and-stick representations, and the production of stereo pairs. It has been used almost universally by small-molecule crystallographers to prepare illustrations of their structures for publication; less-frequently used for proteins.

The Thermal Ellipsoid feature, shown in *Figure 2.3*, requires some explanation. Given sufficient observed data, it is possible to derive from an X-ray crystal structure analysis both the mean position and something about the dynamics of every atom. The dynamical state can be expressed in terms of the amplitudes of vibration along three mutually perpendicular directions. Note that the vibration amplitudes are expected to be anisotropic: it can be seen in *Figure 2.3* that it is more difficult to compress a bond than to rock a bond angle; thus amplitudes of vibration perpendicular to bonds are generally larger than amplitudes of vibration along bonds.

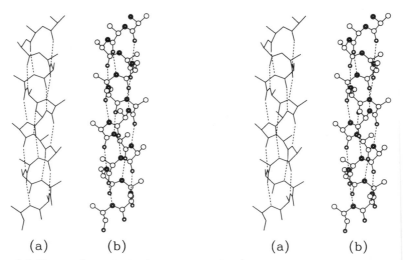

(a)   (b)   (a)   (b)

Figure 2.1 The α-helix, the basis of the structure of hair and wool, and one of the two common structural elements of globular proteins. This structure was predicted by Linus Pauling before its existence in proteins was confirmed. In an α-helix, the N–H group of residue i forms a hydrogen bond to the C=O group of residue $i-4$.

(a) Skeletal drawing. (b) Ball-and-stick model. The convention in this and similar drawings is: carbon atoms are represented by open circles, oxygen atoms by double circles, and nitrogen and phosphorus atoms by triple circles. A polyalanine helix is shown here.

The geometry of the α-helix is as follows: rise per residue 1.5 Å, angular displacement per residue 100 degrees, equivalent to 3.6 residues per turn. The radius of the Cα atom = 2.29 Å, of the Cβ atom = 3.3 Å. A useful 'back-of-the-envelope' calculation: typical helices in a roughly-spherical protein such as myoglobin are up to about 20 residues long. Given that helices tend to go from one end of a molecule to the other, with loops on the surface, it can be deduced that the diameter of myoglobin is about 20×1.5 Å = 50 Å.

Refinement of a full complement of anisotropic thermal parameters as well as the atomic positions requires nine parameters per atom. It is the ratio of the number of observables to the number of parameters that limits the description of atoms in proteins to three positional and a single overall thermal parameter (x, y, z, B) per atom. Moreover, although there is a well-developed theory that provides an interpretation of B-factors on the assumption that they arise exclusively from thermal vibration, and for most small-molecule crystal-structure analyses the B-factors do indeed reflect their true theoretical

Figure 2.2 The β-sheet, the second of the common structural elements of globular proteins. Also predicted by Linus Pauling. β-sheets are formed by the lateral hydrogen-bonding of different strands. The strands may be all parallel (in the sense that the directions from N-terminus to C-terminus are the same), anti-parallel (the strands alternate in direction) or mixed. The structure of silk is based on the anti-parallel β-sheet. Parts (a) and (b) show an ideal parallel β-sheet similar to those found in many proteins, in two different orientations. Note that it is not flat but twisted (21). Part (c) shows an antiparallel β-sheet from the lectin concanavalin A [2CNA]. The second and third strands from the right are joined by a region called a 'hairpin loop'.

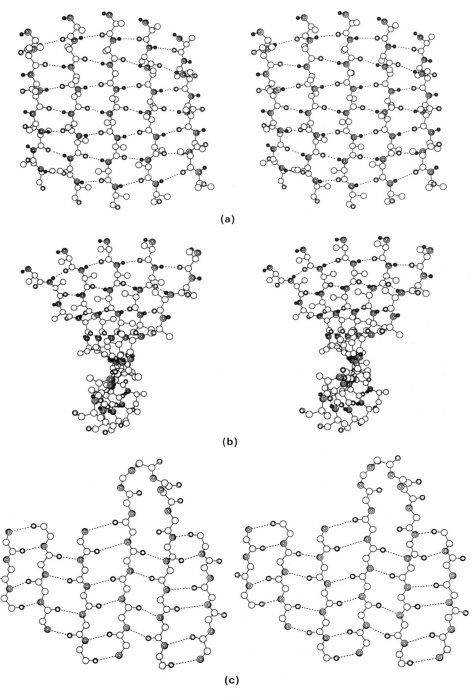

(a)

(b)

(c)

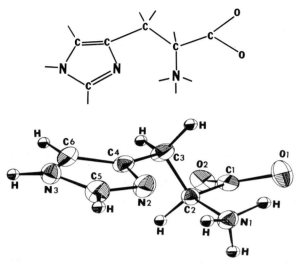

Figure 2.3 A drawing of the molecule histidine produced by C. Johnson's ORTEP (Oak Ridge Thermal Ellipsoid Program). (Courtesy of Dr J. J. Madden.)

significance, in protein crystallography the *B*-factor should be regarded simply as an adjustable parameter reflecting the effective size of the atom (see Section 4.1 in Chapter 1).

The simplest skeletal and ball-and-stick models are easy to produce (see next section). However, in the form of still, black-and-white, pictures they are much less useful for large molecules than for small ones (or for selected parts of large ones: see *Figures 2.1* and *2.2*). Stereo can greatly increase their range of utility (the reader may compare the intelligibility of *Figure 2.1* viewed in mono and in stereo, and this is a *regular* structure). But what really rescued skeletal models of macromolecules was the development of interactive graphics devices that permit the use of real-time rotation, stereo, and depth cueing to help make a complicated drawing understandable (see Section 2.2 in Chapter 1).

In succeeding sections we shall first describe some of the geometric relationships that underlie the transformations from molecular space to viewing screen, and then show some examples of applications to the illustration of biological molecules.

2 From a molecule to a picture

2.1 Introduction

To produce a line drawing using a two-dimensional device such as a pen plotter, film recorder, or laser printer, the programmer must keep two coordinate systems in mind:

The **three-dimensional molecular space** in which the atomic coordinates are given. The units are almost always ångstroms.

The **two-dimensional co-ordinate space** of the page (or screen or film). The units are usually either centimetres or inches (in the case of PostScript* printers the unit is $\frac{1}{72}$ of an inch).

Three steps are required:

1. Decide, within molecular space, what material should appear. This involves selection of regions, decision about which bonds to show, and selection of origin, scale, and orientation. Let us assume here that these problems have been solved, and one has a list of N three dimensional line segments: $(X_i, Y_i, Z_i;$ $X'_i, Y'_i, Z'_i, i = 1, \ldots N)$ such that for each i, (X_i, Y_i, Z_i) and (X'_i, Y'_i, Z'_i) are the end-points of a line segment in molecular space that is to be viewed. The orientation and viewpoint are assumed to be chosen so that the point midway between the viewer's eyes lies on the positive Z-axis.

2. Transform this set of points in molecular space to a set of end-points in two-dimensional image space. This involves a projection on to two dimensions; it may also involve 'clipping' transformations to limit the viewing area, and additional rotations to generate stereo pairs. (Hidden-line removal must *precede* the projection to two dimensions.) This step will produce a set of two-dimensional line segments $(x_j, y_j; x'_j, y'_j, j = 1, M)$—note that M may not equal N—suitable for plotting. These transformations will be the main subject of this section.

   Steps 1 and 2 are purely mathematical, and independent of the equipment available.

3. A program must be available to issue appropriate commands to the driver of the plotting device. The actual form of these commands will depend on the device and the operating system, and will often have local variations superimposed on the manufacturer's protocols. If the reader is attempting to write his or her own program, or to install software written originally for a different device, the best advice I can offer is (strictly in this order) to:

   (a) Find a colleague who has already written a suitable program, and use, copy, or imitate it.

   (b) Failing this, pester the computer centre staff for help.

   (c) RTFM: 'Read the flamin' manual'.

It is clear that a standard set of graphics commands would make it possible to write a more satisfying paragraph at this point. Several such standards have been proposed, but none has yet been universally accepted. One fairly widespread graphics language that is likely to become increasingly useful to readers is PostScript, which can already drive many laser printers, and some workstation screens as well. To draw a 1-inch square, its sides parallel to the edges of the

---

*PostScript is a trade mark of Adobe Systems, Inc.

paper, with its lower left corner 1 inch above the bottom and 1 inch to the right of the left edge of paper, the following file could be sent to an appropriate laser printer (recall that the PostScript unit of length is $\frac{1}{72}$ inch):

```
0.1 setlinewidth
newpath
72      72      moveto
144     72      lineto
144     144     lineto
72      144     lineto
72      72      lineto
stroke
showpage
```

PostScript is, in fact, an extremely rich language for Desk-Top Publishing, with commands for area shading, multiple-font text generation, and page layout that go far beyond its use in this rudimentary example. The reader is referred to the standard reference books (22–24) and two introductory articles (25).

2.2 Geometrical transformations

Suppose that we have the list of line segments in three dimensions, in the proper orientation to be viewed from a point on the positive Z-axis: $(X_i, Y_i, Z_i; X'_i, Y'_i, Z'_i, i=1, \ldots N)$ and we wish to create a list of line segments in two dimensions: $(x_j, y_j; x'_j, y'_j, j=1, \ldots M)$ to plot. The simplest approach is obvious—just throw away the Z coordinate:

$$x_i = X_i \qquad x'_i = X'_i$$
$$y_i = Y_i \qquad y'_i = Y'_i$$

Then scale appropriately.

End of story? Not quite.

This simple approach can be improved in several ways:

- By performing a perspective projection.
- By creating stereo pairs.
- By specifying the spatial limits of the material to be plotted.

If these limits are specified in the viewing plane, one speaks of a 'viewport', and of 'clipping' lines to the viewport. If these limits are specified by bounds to X, Y, and Z in molecular space, one speaks of a 'window'. A special case is a pair of (sectioning) planes perpendicular to the direction of view, in which case one speaks of 'hither' and 'yon' clipping planes. Taking viewport and window definitions together, one has defined a rectangular volume in molecular space to be mapped on to an area of the image plane.

2.2.1 Perspective projections

Perspective was a discovery of Italian artists, notably Brunelleschi, during the Renaissance. It reflects the fact that objects closer to our eyes appear larger, and goes on to specify more precisely the geometrical relationships between the appropriate two-dimensional representation and the three-dimensional space depicted. Consider the image of a (transparent) wire cube, viewed perpendicular to one face. In true geometrical space, the edges parallel to the view direction are truly parallel and their extensions never meet. But because the front face looks larger than the back face, the edges joining them do not appear parallel, and their extensions would appear to meet at a point called the 'vanishing point'.

Let us define the viewing geometry as in *Figure 2.4*, showing a 'side view' of the system, perpendicular to the viewing direction (along Z). The centre of projection is at the origin, which is well outside the volume occupied by the molecule. The eyepoint is at the centre of projection. In front of the eyepoint we construct a 'screen' on which we imagine the image to be projected. The distance from eyepoint to screen is b. The screen can in fact pass through the volume occupied by the molecule; *Figure 2.4* is drawn with the screen between the eye and the molecule only for expository simplicity. Let us further suppose that we have scaled the coordinates so that the volume occupied by the molecule actually has macroscopic dimensions. For example, it would be reasonable to be viewing an object in a 50-cm cubic box, through a 50-cm ($\approx$ 20-in) screen, with an eye-to-screen distance of 50 cm. These numbers approximate the geometry of a scientist sitting at a graphics device.

Calculation of the screen coordinate y_i from the molecular coordinates $P_i = (X_i, Y_i, Z_i)$ is a simple exercise in 'similar triangles', where

$$\frac{y_i}{b} = \frac{Y_i}{Z_i}$$

and an analogous ratio defines x_i.

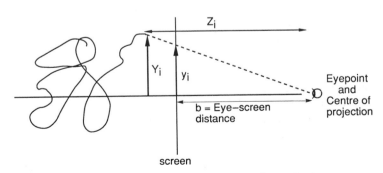

Figure 2.4 Viewing geometry for perspective projection.

25

Solving these:

$$x_i = b\,\frac{X_i}{Z_i}$$

$$y_i = b\,\frac{Y_i}{Z_i}.$$

Let us compare the effect of this transformation on two points with the same X and Y coordinates but different Z coordinates: (X, Y, Z_1) and (X, Y, Z_2), with corresponding screen points $(bX/Z_1, bY/Z_1)$ and $(bX/Z_2, bY/Z_2)$. If $Z_1 > Z_2$, then $bX/Z_1 < bX/Z_2$ and $bY/Z_1 < bY/Z_2$; thus objects more distant from the eye: $Z_1 > Z_2$, will indeed appear smaller: $bX/Z_1 < bX/Z_2$.

If we keep the screen size and the eye-to-screen distance fixed, and move the object back away from the screen, two things will happen:

1. As the distance from the molecule to the point of projection becomes much larger than the size of the molecule, the foreshortening will decrease, and in the limit the projection will approach the simple orthogonal projection:

   molecular space $(X, Y, Z) \rightarrow$ screen space (X, Y).

   The reason is that if the distance from the eye to the object is much larger than the range in depth of the object, the Z coordinates of the points in the object become effectively constant.

2. But, the size of the image will grow smaller and smaller on the screen. We can always scale it up; indeed, we can always arrange for the object to fill the screen. It is quite interesting to go through the algebra (readers who do not believe this can skip over it) because it shows that if we continually rescale the object as it moves towards us or away from us, we will retain the information that perspective can provide about relationships within the object, but we will lose the information that perspective can provide about its distance from our eye. This implies that automatic scaling will be useful for single or 'still' pictures, but would be inappropriate for an interactive graphics session with manipulative control over an object's position in space.

Referring to *Figure 2.5*, suppose that the object we wish to display is contained in a cube of edge S, with its centre at $Z = Z_v$. We have chosen S so that the complete cube maps into the full extent of a square screen of height a located at a distance b from the eyepoint. Note that given a, b, and Z_v, the size of S is determined by the conditions that the corners of S nearest the eyepoint map into the corners of the screen.

These conditions require that:

$$\tfrac{1}{2}S, \tfrac{1}{2}S, Z_v - \tfrac{1}{2}S \text{ maps into } (\tfrac{1}{2}a, \tfrac{1}{2}a),$$

that is:

$$\tfrac{1}{2}a = b\,\frac{\tfrac{1}{2}S}{Z_v - \tfrac{1}{2}S}$$

26

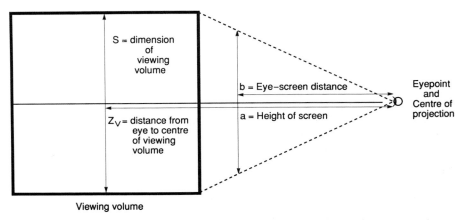

Viewing volume

Figure 2.5 Scaling so that viewing volume to fill the screen.

or:

$$S = \frac{2Z_v}{1 + 2(b/a)}.$$

(A quick check: if the front face of the cube is at the screen: $Z_v - \frac{1}{2}S = b$, then $S = a$.)

In words: if b and a are fixed, the largest cube centred at Z_v that will be completely visible will have edge

$$S = \frac{2Z_v}{1 + 2(b/a)}.$$

To apply this result to scaling of molecular coordinates, let us redefine the origin of molecular coordinates to be at the centre of the cube rather than at the eyepoint:

$$X_i'' = X_i$$
$$Y_i'' = Y_i$$
$$Z_i'' = Z_i - Z_v.$$

If all molecular coordinates (X_i'', Y_i'', Z_i'') fit into a cube of edge S'' centred at $Z'' = 0$, (that is, $Z = Z_v$), we know that we can scale them by the factor

$$\frac{S}{S''} = \frac{\dfrac{2Z_v}{S''}}{1 + 2(b/a)}$$

27

to map them into the largest cube centred at Z_v that will map into the entire screen. Then the scaled screen coordinates are given by:

$$x_i = b \, \frac{\dfrac{S}{S''} X_i''}{\dfrac{S}{S''} Z_i'' + Z_v} = b \, \frac{X_i''}{Z_i'' + \dfrac{S''}{S} Z_v}$$

$$y_i = b \, \frac{\dfrac{S}{S''} Y_i''}{\dfrac{S}{S''} Z_i'' + Z_v} = b \, \frac{Y_i''}{Z_i'' + \dfrac{S''}{S} Z_v} .$$

But $\dfrac{S''}{S} Z_v = \tfrac{1}{2} S''[1 + 2(b/a)]$;

therefore we can even go one step further and make the results independent of the size of the screen, by calculating the ratios x_i/a and y_i/a (fractional screen coordinates):

$$\frac{x_i}{a} = \frac{b}{a} \frac{X_i''}{Z_i'' + \tfrac{1}{2} S_i''[1 + 2(b/a)]}$$

$$\frac{y_i}{a} = \frac{b}{a} \frac{Y_i''}{Z_i'' + \tfrac{1}{2} S_i''[1 + 2(b/a)]} .$$

It is most interesting that the result now depends only on the ratio b/a, and is independent of Z_v: if we move the object back it will automatically be scaled up!

A very important point to note about these equations is that we do not have to worry at all about the relationship between microscopic and macroscopic units. It is only necessary that we measure the molecular coordinates X'', Y'', Z'' and their range S'' in consistent units (ångstroms are the obvious choice), and that we measure b and a in any (other) consistent units (centimetres, inches, or any other choice).

These equations imply that the perspective distortion of the object depends only on the ratio of the quantities b and a which characterize the geometry of the viewing system. By varying the ratio a/b we generate images with different degrees of foreshortening. *Figure 2.6* shows some examples.

In contrast, if we do the original mapping without rescaling, the picture will change as we move the object towards us or away from us: If we keep the position of the object fixed, but scale the molecule within the viewing volume (relative to an origin within the molecule, not at the eyepoint), we will generate images with the same degree of foreshortening, but having different sizes on the screen. If we reduce the scale, the image will occupy only a part of the screen. If we increase the scale, the lines may go outside the screen area. (For some combinations of sizes

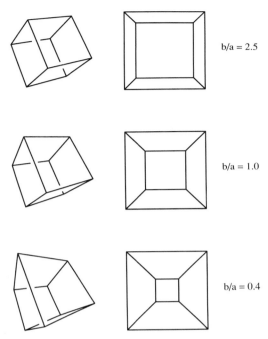

Figure 2.6 Drawings of a cube showing different extents of foreshortening.

and positions, some of the atoms may even be behind the eyepoint! What we must do then is the subject of Section 2.2.4.)

For additional information about this subject, see refs 5, 26, and 27.

2.2.2 Stereoscopic views

The generation of right and left stereo views is an extension of the techniques described in the previous section.

We have already introduced the basic idea that the task is to create two images of an object that correspond to right-eye and left-eye views (see Section 2.2.1 of Chapter 1). The basic geometrical relationships illustrated in *Figures 2.4* and *2.5* are still applicable. In principle, the generation of a left-eye view involves resetting the eyepoint by a small translation. This is, of course, equivalent to translating the object by the same amount in the opposite direction. After projection the image can be recentred on the screen. In practice, it is even easier to approximate this effect by rotating the object by ±3 degrees (the exact value is not critical) and doing the standard projection. In most cases the visual system can accommodate, especially if the range in depth of the atoms is not too large.

Of course the *presentation* of the two views so created separately to the two eyes is another problem which usually requires special equipment.

However, there are some geometrical aspects of the situation worth noting.

Whereas for single (monocular) pictures it is unnecessary to worry about the relationship between microscopic and macroscopic units, for stereo drawings certain macroscopic distances are fixed, by anatomical and physiological constraints, or by the viewing equipment. The anatomical constraint is the interocular separation, typically about 6.5 cm in the human adult. The physiological constraint is the effective range of stereoscopic vision, which is (on the average) optimal at an effective viewing distance of about 60 cm.

The following geometrical parameters for creating stereo images are based on those suggested by Johnson (12): The image of the molecule occupies a box, 12 inches square in directions perpendicular to the viewing direction, and 6 inches deep (the box is relatively shallow to minimize the inaccuracies introduced by approximating translation by rotation in creating the two views). The plane of projection is half-way back into the box. The distance from the centre of the box to the point midway between the eyes is 30 inches. Note that this fixes the ratio b/a (see *Figure 2.5*) at 2.5 (cf. *Figure 2.6*). To create the separate images, Johnson suggests either translations of ± 3.25 cm (half the interocular separation) or rotations of ± 3 degrees.

Although an accurate stereoscopic picture requires considerable attention to the geometry of the viewing system for which it is intended, in practice the parameters in common use give acceptable results.

2.2.3 Window–viewport mapping; clipping

Suppose we wish to project a specified volume in molecular space (the window) on to a specified area (the viewport) on a screen. For simplicity let us assume that the volume is a rectangular solid and the area is a rectangle, both with axes parallel to the coordinate axes (Generalizations involve technical difficulties rather than new principles: circular viewports can be useful—see Chapter 4—and viewports shaped like the windows of cockpits in real aircraft have been used in pilot-training simulation systems.) If a line segment lies entirely outside the region, it will be deleted entirely. If a line segment crosses a boundary, we want to truncate or 'clip' it to that portion within the boundary. *Figure 2.7* shows examples of clipping in two dimensions. The three-dimensional windowing— that is, the restriction of the lines to the three-dimensional window—must be done before projection (this is obvious) and the clipping must be done after projection (slightly less obvious).

The purpose of this section is not to describe clipping algorithms, which are the subject of a substantial technical literature, but to introduce the reader to some of the relevant terminology.

Recall that a 'window' is a three-dimensional rectangular volume in molecular space oriented along the coordinate axes, that defines the range of line segments that may be drawn.* The two planes that bound this volume perpendicular to the

*The word 'window' is used in several different ways in computing. This definition is consistent with the Evans & Sutherland PS-300 language definition.

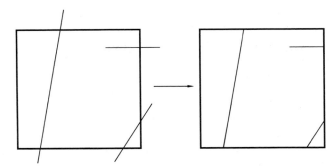

Figure 2.7 Clipping in two dimensions.

view direction, nearer to and farther from the eyepoint, are called the 'hither' and 'yon' clipping planes.

The portions of line segments lying within the window volume are projected on to a plane as in *Figure 2.4*. Of course, if the window is made larger than the extent of the molecule, no clipping will occur. The extreme points of the window volume map into a rectangle in the projection plane; let us call this the projection of the window: note that it is a two-dimensional rectangular region. Another rectangular region, the viewport, specifies the area on the physical screen or paper in which the image is to appear. To produce the picture, the projection of the window is mapped on to the viewport (a mapping between two two-dimensional regions). The ratio of length to width of a rectangle is called its 'aspect ratio'. If the projection of the window has a different aspect ratio from the viewport, the mapping will alter the shape of an object: for example, if the projection of the window is square but the viewport is short and wide, a sunfish will appear on the screen as an eel.

To show left and right stereo images side-by-side, a common window would be mapped (via slightly different transformations) into two alternative viewports, displaced horizontally. To show left and right stereo images separated by alternation in time, both can be mapped into the same viewport.

2.2.4 Summary

The steps by which a picture of a molecule is brought to visibility on a screen or piece of paper may be recapitulated:

1. The coordinates are scaled in size and oriented relative to a three-dimensional volume of visibility (the window). If two stereo views are desired, the two required positions or orientations are generated, and steps 2 and 3 are carried out in parallel on the two views.

2. The line segments are clipped, if necessary, so that segments or portions of segments outside the window are deleted.

3. The remaining material is mapped on to a plane of projection, according to the viewing geometry specified.

4. The rectangle corresponding to the extreme points of the window is mapped linearly on to the viewport area of the screen or paper. If two stereo views are to be shown side-by-side, the window will be mapped on to two separate displaced viewports.

On many modern interactive graphics devices, all these transformations are carried out by hardware, at interactive speeds, or, in jargon, in 'real time'. For most plotters and ordinary graphics terminals, systems software exists to carry out at least some of the steps. It is unlikely that anyone who reads this book will ever have to write programs, from scratch, to carry out these operations. But a certain degree of understanding of what the hardware or software is doing will contribute to one's being able to use it intelligently and effectively.

2.3 Creating the picture from the coordinates

In the last section, we proceeded from the assumption that one had a set of three-dimensional line segments, corresponding to a drawing of a molecule, and described how it is brought to the screen. Here we shall discuss how, starting with a set of coordinates, one can generate such a drawing. In this chapter we will treat skeletal and ball-and-stick drawings.

2.3.1 Skeletal drawings

In this section we consider simple pictures in which there is a direct correspondence: one bond—one line segment. Given a set of atomic coordinates (for example, a Protein Data Bank file) how can we extract the bonds? There are basically two ways:

1. Screen the atoms by distance. This is the most general approach. For every pair of atoms in the structure, calculate the distance between them. If the distance is less than the sum of the Van der Waals radii of the two atoms, assume that there is a bond between them. For proteins, this approach can be specialized by checking only atoms in the same residue, plus the atoms in the peptide bonds between successive residues. Similar considerations apply to nucleic acids.

   Distance screening can obviously fail in two ways, especially when errors in the coordinates are large. One may miss bonds if two atoms are too far apart, or draw spurious bonds if two non-bonded atoms appear too close together. For this reason, one may prefer to:

2. Create an explicit list of bonded pairs. For proteins containing only standard amino acids and common ligands, this can be done once and for all. For each residue type one can make a list of pairs of atom names, each pair corresponding to a bond. Then in drawing a picture of a protein, for each

residue one can search for the coordinates of each pair of atoms in the list of bonds and add the appropriate line segment to the drawing. It is necessary to check explicitly for breaks in the chain before blindly connecting successive residues with peptide bonds.

For non-standard groups, one may either construct a connectivity list by hand, or fall back on distance screening. In Protein Data Bank files, explicit connectivity lists are provided for non-protein moieties.

2.3.2 From bonds to balls-and-sticks

A ball-and-stick drawing may be considered a development of a simple skeletal model, in which the representation of the bond is generalized to a 'cylinder' and a disc is added at the position of each atom. Additional information may thereby be displayed, in that different atom types may be distinguished by size and shading, and bonds of different appearance may be drawn—for example, covalent bonds may be shown as thick solid lines, and hydrogen bonds as thin dashed ones.

To create the line segments corresponding to a pure skeletal model is simple: one need only copy the coordinates of each pair of bonded atoms as the line segment end-points. To create ball-and-stick pictures, one must:

(a) draw a circle—possibly a filled circle or a shaded sphere—at the position of each atom. It is useful to have the facility to vary the atomic radii; and

(b) determine the line segments that represent the bond. Unless one is making use of a general hidden-line removal program, it is useful to 'clip' these segments at the edge of the circle.

Drawing a circle sounds (and is) trivial. One stores in a table the values of

$$\left(\sin \frac{2\pi}{n} k, \cos \frac{2\pi}{n} k \right), \quad k = 1, \ldots n,$$

and when the time comes to draw a circle of radius r centred at a point x, y, z, one executes code to:

$$\text{move to } x + r, y, z$$

$$\text{for } k = 1, \ldots n: \text{ draw to } x + r \cos \frac{2\pi k}{n}, y + r \sin \frac{2\pi k}{n}, z.$$

To generate the segments representing the bond, let us assume that the two atoms have coordinates $v_1 = (x_1, y_1, z_1)$ and $v_2 = (x_2, y_2, z_2)$ and radii r_1 and r_2, respectively. We wish to represent the bond by n line segments on the surface of a cylinder of radius r of which the line $v_1 - v_2$ between the atoms is the axis. Assume that r_1 and r_2 are both $> r$, and that n is even.

If $n = 2$, the line segments define the silhouette of the cylinder, and we should like the line segments separated by r to give an 'open' bond. To generate the direction perpendicular to both the view direction and the inter-atomic vector, calculate the cross-product $w = z \times (v_2 - v_1)$, where z is a unit vector along the positive z-axis, assumed to be the axis of view.

$$\text{Scale this to length } r: p = r \frac{w}{|w|}.$$

Then the end-points of one of the line segments in the representation of the bond are the intersections of the line segment from the point $v_1 + p$ to the point $v_2 + p$, with the spheres around the atoms. Referring to *Figure 2.8*, we see that a length $\sqrt{(r_1^2 - r^2)}$ must be clipped off the end of the segment at atom v_1, and similarly a length $\sqrt{(r_2^2 - r^2)}$ must be clipped from the other end. Therefore the two end-points of the desired line segments are:

$$v_1 + p - \Delta v \sqrt{(r_1^2 - r^2)}$$

and

$$v_2 + p - \Delta v \sqrt{(r_2^2 - r^2)},$$

where $\Delta v = (v_1 - v_2)/|v_1 - v_2|$.

If $n \geq 2$, other segments on the cylinder representing the bond can be generated by rotating these points around the vector $v_1 - v_2$. If we let

$$g_k = \left(\cos \frac{2\pi}{n} k \right) p + \left(\sin \frac{2\pi}{n} k \right) (p \times \Delta v),$$

then the segment end-points are:

$$v_1 - \sqrt{(r_1^2 - r^2)} \Delta v + g_k, \; v_2 + \sqrt{(r_2^2 - r^2)} \Delta v + g_k, \; k = 1, \ldots n.$$

Additional complications are introduced if it is desired to taper the bonds away from the viewer (to give an exaggerated 'perspective' effect). These will not be treated here.

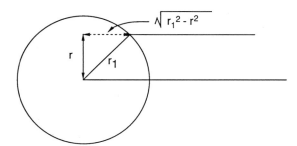

Figure 2.8 Clipping a stick to a ball.

3 Applications

Wire models and ball-and-stick models are extremely useful because of the great detail they contain. For still pictures, it is necessary that one is dealing with a sufficiently small molecule—or a small enough part of a large molecule—for them to be intelligible. (They are particularly useful in connection with 'blow-ups' of selected portions of a large molecule, to be discussed in Chapter 4.) To push the intelligibility of a fully-detailed still picture to larger molecules, stereo is essential.

These warnings apply with somewhat less stringency to pictures produced with high-performance interactive graphics devices, because the ability to view the figure in motion greatly enhances its intelligibility.

The examples that follow are intended to illustrate some of the strengths and weaknesses of these representations, permitting the reader to judge for himself or herself when they might be useful.

3.1 The components of nucleic acids

Figure 2.9, showing the standard Watson–Crick base-pairs, illustrates molecules small enough for a fully detailed picture to be entirely appropriate. (The next chapter contains pictures of the amino acids.)

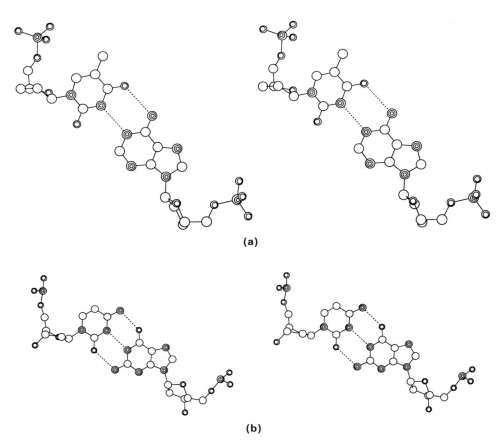

(a)

(b)

Figure 2.9 Watson–Crick base-pairs. (a) adenine–thymine. (b) guanine–cytosine.

3.2 Small proteins

How large a molecule can be usefully represented in a fully-detailed picture? *Figures 2.10–2.12* contain a series of pictures of proteins of increasing size, in both skeletal and ball-and-stick representations: Glucagon contains 29 residues and

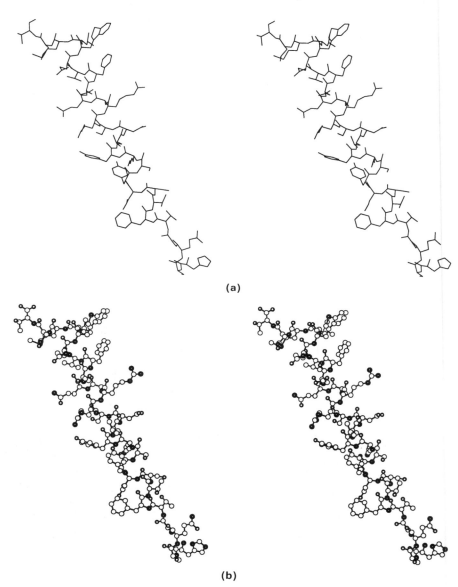

(a)

(b)

Figure 2.10 Glucagon [1GCN] 29 residues.

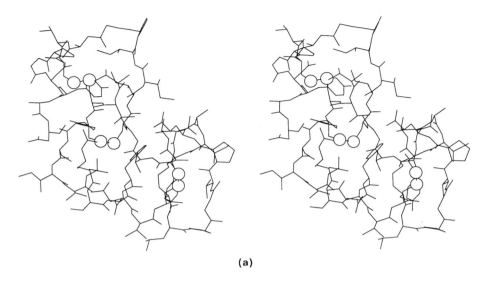

(a)

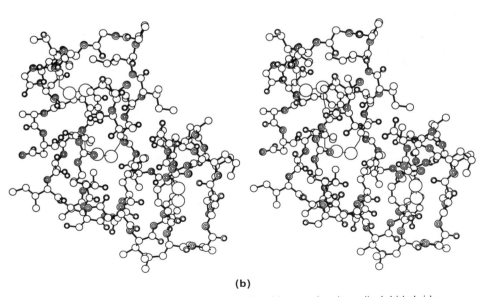

(b)

Figure 2.11 Crambin [1CRN] 46 residues. Crambin contains three disulphide bridges.

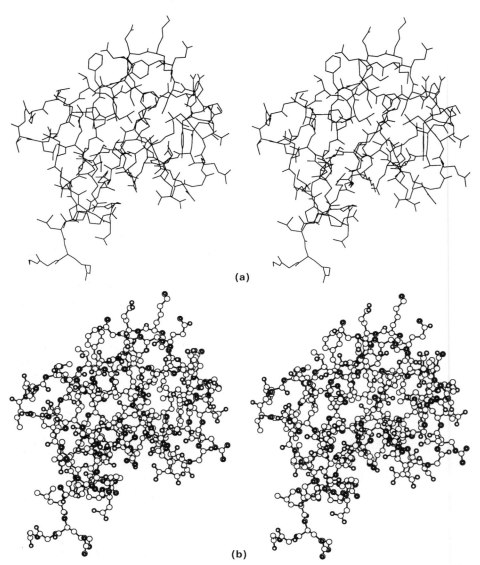

(a)

(b)

Figure 2.12 Ubiquitin [1UBQ] 76 residues

246 non-hydrogen atoms; crambin contains 46 residues and 327 non-hydrogen atoms; ubiquitin contains 76 residues and 602 non-hydrogen atoms. The reader is urged to look at these, first in mono and then in stereo, to try to decide at what point the molecules are too big for these representations to be useful. Which representation can be 'pushed further'; that is, which remains intelligible for larger structures?

3

Shaded-sphere pictures and the packing of atoms in proteins

1 Introduction

In this chapter we discuss the second of the traditional representations of molecules in which complete chemical detail is represented. In this case it is each atom rather than each bond that is shown. This category includes three basic types of pictures:

1. Line drawings in which each atom is represented as a disc. We have already mentioned this possibility as a limit of the ball-and-stick drawings as the radius of each atomic 'ball' is made as large as the Van der Waals radius. However, we have not described in detail how to solve the hidden-line problems that arise both as bonded atoms intersect and as atoms on different levels occlude each other. (See *Figure 3.1*.) A special case of this representation is the cutting of serial sections through a protein; F. M. Richards developed

Figure 3.1 Picture of a haem group in which each atom is represented as a disc with radius equal to its Van der Waals radius.

39

this method and showed the significance of the packing of protein interiors that it reveals. (See *Figure 3.2.*)

2. Using colour raster devices, it is possible to draw each atom as a shaded sphere, or even to simulate the appearance of the well-known Corey–Pauling–Koltun (CPK) physical models. With the usual facilities of an 8-bit per pixel screen, it is possible to maintain most of the familiar colour scheme (C = black, N = blue, O = red . . .), except for the great difficulty of computing the appearance of the black carbon atoms.

In such a representation, atoms are usually opaque, so that only the front layer of atoms is visible. However, clipping with a 'hither' plane (a 'cheese-wire cut') can show the packing in the molecular interior. (See *Plate 1*.) Until recently, raster devices were not powerful enough to achieve real-time rotation of a sizeable shaded-sphere picture. The still pictures to which one was thereby restricted suffered, as do wire models, from the relative lack of intelligibility of a picture of a large molecule. However, with the introduction of the new generation of 'superworkstations' it is possible to rotate images such as *Plate 1* in real time, with the enhancement of the perception of spatial relationships that this entails and, perhaps of even greater significance, the possibility of moving the hither clipping plane through the molecule in real time, to show the packing in the protein interior.

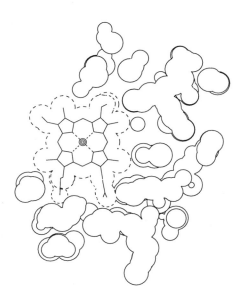

Figure 3.2 Three serial sections through cytochrome c [4CYT] showing packing around the haem group. Van der Waals contours around the haem group are shown in broken lines; Van der Waals contours around other atoms are shown in solid lines. The haem group is shown as a wire model, with the iron at its centre. The frontmost slice is in the mean plane of the haem group.

3. The real-time rotation and clipping facilities available on vector graphics devices are very useful in connection with another technique for representing atomic and molecular surfaces. The 'spatter-painting' of the surface of a sphere by a distribution of several hundred dots requires many fewer dots than would be required to represent the complete surface of the sphere, and it produces a 'translucent' representation of the surface. (See *Figure 3.3*.)

It is entirely possible to combine dot-surface representations with skeletal models to show both the topology of the molecule and its space-filling properties. M. Connolly (28, 29) has developed this technique to show, not the surface of the union of Van der Waals spheres, but the much smoother solvent-accessible surface. [The solvent-accessible surface of a molecule was defined originally by Lee and Richards (30) as the surface generated by the centre of a sphere of the radius of a water molecule (1.4 Å) rolled around the protein. It is smoother than the Van der Waals surfaces of the atoms, because the Van der Waals surface contains small crevices that water cannot reach.] *Plates 2* and *3* compare the Van der Waals and Connolly surfaces.

Although the molecular illustrator R. Hayward drew by hand many pictures of small molecules that are related to this representation, colour graphics with real-time rotation and other advanced features were necessary to unlock the real power of the technique.

2 Shaded-sphere pictures on raster devices

2.1 The nature and operation of raster devices

A raster device is a specialized piece of hardware, equipped with a screen, that can map an array stored in memory on to the screen so that the value of each element of the array controls the appearance of the corresponding point on the screen. Raster devices vary in the number of points on the screen (1024×1024 is the typical size*) and in the amount of information that corresponds to each pixel. 1 bit per pixel corresponds to a black-and-white monitor on which each point is either light or dark; 8 bits = 1 byte per pixel is quite common with colour monitors, giving a total of 256 colour-intensity levels. The number of bits per point is called the 'number of bit planes'. With an 8-bit plane monitor, the user has the option of assigning colour-intensity combinations to each of the 256 possible values. The monitor of such a device will have three colour 'guns'—Red, Green, Blue—the intensities of which can be varied independently.

Suppose one has a 1000×1000 point monitor on an 8-bit-plane device. To create and display a picture, a program must make up a 1000×1000 byte array by calculating the appropriate value for each element. This calculation is done with reference to a video-look-up-table (common abbreviations: VLT or LUT), which

---

* In many devices the visible area is 1024 (horizontal) $\times N$ (vertical) where N is somewhat smaller than 1024. This is a consequence of the capacity of the monitor.

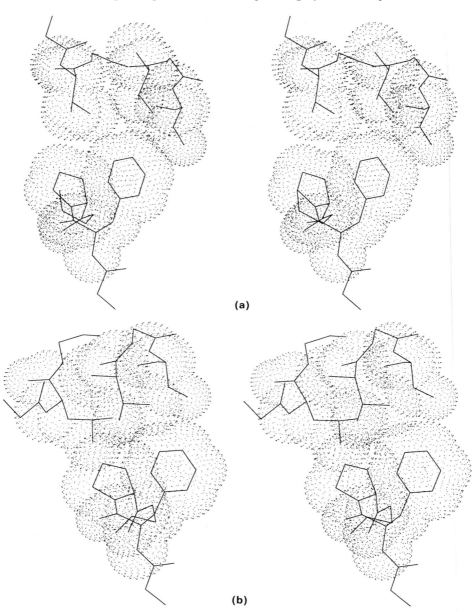

(a)

(b)

Figure 3.3 A molecular 'ball-and-socket' joint at the interface between the V_H and C_{H1} domains of immunoglobulins. In both cases the 'ball' is formed by a Phe and Pro from the C_{H1} domain (lower part of each picture) and the 'socket' by a Val, Ser, and Thr from the V_H domain (upper part of each picture). The rotation around this joint is involved in changes in the 'elbow angle', the angle between the V_L-V_H domains and the C_L-C_{H1} domains. (a) Fab KOL [1FB4], (b) Fab McPC603 [1MCP]. See ref. 31.

assigns to each of the 256 possible pixel values a set of three intensities corresponding to the Red, Green, and Blue components of the image. A programmer who wanted to display 8 colours, each with 32 intensity levels, would make up a suitable VLT and compute pixel values accordingly.

The video look-up table could be a 3×256 element array with values as follows:

VLT

Intensity of:

Element number	Red	Green	Blue	
0	8	0	0	weak red
1	16	0	0	
2	32	0	0	
...		...		...
30	248	0	0	
31	255	0	0	bright red
32	0	8	0	weak green
...		...		...
63	0	255	0	bright green
64	0	0	8	weak blue
...		...		...
95	0	0	255	bright blue
96	8	0	8	weak magenta
...		...		...
127	255	0	255	bright magenta
...				

In the array stored in memory, any point in a 'red' object (e.g. an oxygen atom) would be given a value from 0–31, according to its computed intensity. A point in a 'blue' object (e.g. a nitrogen atom) would be given a value between 64 and 95, according to its computed intensity. Obviously, more complex video-look-up tables can be constructed; in particular, any desired hue can be expressed as a suitable combination of the three primary colours. This limitation to a discrete set of intensity levels can lead to discontinuities in the shading of a surface. Television sets achieve fully-realistic colour ranges, but they are not limited to a 256-element VLT. The best raster graphics devices support many more bits per pixel and allow more intensity levels per colour.

The VLT is quite short, but a full screen on a 1000×1000 8-bit plane device contains a megabyte of information. Transmission of a megabyte of data requires about 10 sec at ethernet speeds;* it is this bottleneck that makes it impossible to

---

* Ethernet is a system for local communication between computers. Information transfer rates are relatively fast, but limited in distance; therefore ethernet links are typically useful within a single building. A new protocol, designed around fibre optic connections, is likely to be 10 times faster.

animate images in a general way on raster devices coupled loosely to a host computer.

The passive terminals described in the preceding paragraph should be distinguished from such devices as the Evans & Sutherland PS-390, which combines vector and raster graphics with many special and powerful hardware features, and from the new generation of graphics workstations. These devices are equipped with very-high-capacity channels for loading pictures on to their screens.

2.1.1 Colour perception and the numerical specification of colour

How are colours classified and specified quantitatively? If one notices an attractive colour, how can it be reproduced in a computer-generated picture?

These questions are part of the very complex topic of colour perception, which combines problems from the 'P*Y' sciences: PhYsics, PhYsiology and PsYchology. This section contains minimal, pragmatic answers.

Fundamentally, we are concerned with the correspondence between:

Spectrum of light entering the eye (that is, intensity as a function of wavelength)→a perception of a colour.

Basic observations about this correspondence are:

1. Any colour can be matched by a suitable combination of three primary colours (for example, red, green, and blue). The observed colour of an object (observed in isolation against a black background) depends on the spectrum of the light it emits, transmits, or reflects; however, two objects with different spectra may be perceptually equivalent.

2. If spectrum A and spectrum B match perceptually in colour, and spectrum C and spectrum D also match, then the superpositions (spectrum A + spectrum B) and (spectrum C + spectrum D) also match perceptually.

   Observable colours may be distinguished on the basis of three characteristics:

3. Hue: 'colour' in the most common colloquial sense (red, green, and blue describe different hues); for monochromatic light, different wavelengths correspond to different hues, but in general different spectra can give the same perceived hue. A flat spectrum: $I(\lambda) =$ constant independent of wavelength, appears achromatic: white, grey, or black.

4. Tone (or value): roughly speaking, the total amount of light per unit area. Multiplying a spectrum by a constant (independent of wavelength) changes the tone. Members of the series white→grey→black differ in tone. Because we are dealing with physiology as well as physics, empirical scales of tone are not linear in integrated intensity; and moreover changes in tone can alter perceived hue.

5. Saturation (or intensity or chroma): The difference between a colour and the grey with the same tone. A pure or saturated colour can be diminished in

saturation by adding white light and normalizing the result to the same perceptual tone.

As a result, colours may be thought of as points in a three-dimensional space, the axes of which might be the primary colours red, green, and blue (or cyan, magenta, and yellow). Each colour is a vector the components of which are the intensities of the primaries (e.g. red, green, and blue) required to match it. For displays generated by three-gun monitors, these are the numbers we wish to specify.

Parallel vectors correspond to colours of the same hue but different tone. Therefore we can specify hues by normalized vectors in this R-G-B space. The choice of the basis vectors in this space must be defined by convention. The Commission Internationale de l'Éclairage (CIE) has chosen primaries to be monochromatic stimuli at 700.0 nm (red), 546.1 nm (green), 435.8 nm (blue) (see ref. 32).

Astrua (33) and Kueppers (34) illustrate the colours arising from mixing different combinations of primaries. These may be used to prepare a numerical palette for designing the colour scheme of a picture. It is possible to choose other primaries; which can, in fact, be any spectral distributions (that are linearly independent in the perceptual sense), not just monochromatic stimuli. The components of a colour expressed in terms of a more general basis set are called tristimulus values (X, Y, Z). In 1964 the CIE chose a specific set of spectra for such a basis. In industrial practice, these or the Munsell system are often used to specify colours. The conversion between tristimulus values (X, Y, Z) and primary components (R, G, B) is given by the following transformation:

$$X = 2.3646\ R - 0.5151\ G + 0.0052\ B$$
$$Y = -0.8965\ R + 1.4264\ G - 0.0144\ B$$
$$Z = -0.4681\ R + 0.0887\ G + 1.0092\ B$$

2.2 Generation of the picture

Now that we know how to display a picture, how do we compute the colours and intensities that correspond to a collection of spheres?

Let us begin with a single sphere. We may assign the colour according to atom type—typically to simulate the CPK colour scheme—or by residue number: often it is useful to distinguish an important set of residues by special colours. When making cheese-wire cuts, it may be helpful to colour the surfaces of truncated atoms by stippling of different colours (*see Plate 1*).

2.2.1 Reflectance models

Suppose that we have a single complete sphere (only the front hemisphere is visible of course) and wish to draw it in some uniform colour but with the surface 'shaded' to appear realistically three-dimensional. The intensity I corresponding to any point on the sphere depends on the reflectance model we take for the

45

surface, and the nature and position of the assumed light source(s). The shading model may include the following contributions:

1. Ambient light; constant background $I_a = c_a$. (In these formulae, c_a, c_d and c_s are constants and I_a, I_d, and I_s stand for contributions to the intensity.)

2. Diffuse reflection: the light is reflected equally in all directions (unlike from a mirror) so that the intensity depends on the angle θ between the direction of the light source and the normal to the sphere at the point: $I_d = c_d I_0 \cos \theta$, where c_d is a reflectance constant, and I_0 is the intensity of the incident beam.

3. Specular (or mirror-like) reflection. The light is reflected only in a special direction, for which the angle of incidence and angle of reflection are equal. The reflected intensity from a point into the viewing direction depends on the angles between the light source, the normal to the sphere, and the viewing direction. It is this effect that puts the highlight on a bright, shiny apple, which, if inspected closely, is under typical illumination bright and shiny only in a small region. Portrait painters place a point highlight on the images of their subjects' eyeballs. A simple mathematical model for specular reflection, developed by Phong, gives:

$$I_s = c_s I_0 \cos^n \psi,$$

where ψ is the angle between the reflected ray and the viewing direction. Increasing the value of n narrows the distribution of reflected light (sharpens the highlight). Thus high values of n give the appearance of metallic surfaces; small values give the appearance of less-shiny materials. (Although this equation is appropriate for monochrome pictures, for truly realistic colour pictures it must be modified in view of the fact that the spectral distribution of reflected light is unequal to that of the incident beam.) Combining these effects:

$$I = I_a + I_d + I_s$$
$$= c_a + c_d I_0 \cos \theta + c_s I_0 \cos^n \psi.$$

In the case of atoms, it is appropriate to regard c_a, c_d, c_s, and n as purely empirical parameters which the user may vary to produce any effect he or she deems pleasing.

The literature of computer graphics reveals considerable efforts to refine this simple model in order to achieve truly convincing representations of real objects like cars and airplanes. These include more accurate treatment of the spectral distribution of reflected light, shadows, multiple reflections, translucency, and the modelling of irregular (e.g. dirty or cracked) surfaces. The interested reader may consult ref. 9. It makes our life somewhat more convenient not to have an objective test of the appearance of molecular pictures.*

---

* Many centuries ago, a Chinese artist painting for the King of Ch'i was asked by the King which was harder to draw: dogs and horses, or demons and goblins. 'Demons and goblins are much easier, Your Majesty', he replied, 'for everyone knows what dogs and horses look like.'

The most general way to draw a sphere [of radius r_0, centred at (x_0, y_0, z_0)] on the screen of a raster device is to choose a set of points on the front surface of the sphere: x_i. y_i, $z_i = [r_0^2 - (x_i - x_0)^2 - (y_i - y_0)^2]^{1/2}$, use the formulas just presented to calculate from a specified lighting/reflectance model the colour and intensity of the point, and map it on to the screen. The points may be spaced equally in x and y, and chosen densely enough so that there is a point corresponding to every pixel in the image of the sphere.

Many raster devices permit drawing a single filled circle of constant colour and intensity in a single operation. With these devices a sphere can be built up from back to front by drawing successive circles—of the same colour, decreasing radius, and increasing intensity—to give the appearance of a sphere lit from the front.

At least one new graphics device (manufactured by the Stardent Computer Corporation) supports the drawing of a shaded sphere as a primitive operation.

2.2.2 Assemblies of spheres

Now that we know how to draw a single sphere, how can we draw a whole molecule, ensuring that we will not overwrite an atom on the front surface by another atom behind it?

For spheres, there is a particularly simple solution: calculate the 'leading edge' of each sphere: $z_i + R_i$ where z_i is the z-coordinate of the centre of the sphere, R_i its radius, and we assume that the molecule is near the origin and that the viewing point is far out along the positive z-axis. Then we sort the spheres in order of leading edge, and draw the farthest away first, so that any point will be overwritten only by a point from another sphere nearer the eye.

This approach depends on the special geometric properties of spheres. A more general approach would be to maintain two arrays, one holding the values of the elements of the picture itself, $P(i, j)$, and the corresponding entry in the second, $B(i, j)$, recording the z-coordinate of the point that contributed the value of each pixel to the current image. Then we can draw all objects in a scene, and let points nearer the eye 'overwrite' points farther from the eye whenever they map into the same point on the screen. Whenever a surface point (X, Y, Z) maps into a picture element $P(i, j)$, the corresponding pixel value is discarded if $B(i, j) < Z$. If $B(i, j) \geqslant Z$, *both* $P(i, j)$ and $B(i, j)$ are updated. This is called the Z-Buffer technique. (To represent translucency another buffer is required.)

2.3 Examples: The amino acids

For comparison, *Figure 3.4* shows the structures of the twenty amino acids that are the building blocks of proteins, in both shaded-sphere and ball-and-stick representations.

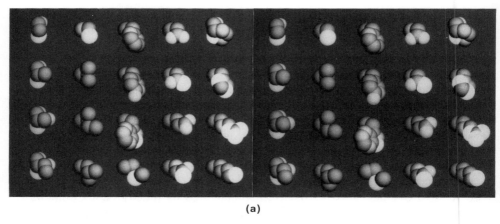

(a)

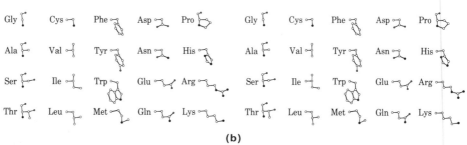

(b)

Figure 3.4 The structures of the twenty amino acids that make up proteins. (a) Space-filling models, (b) ball-and-stick models. The images of Gly, Ala, Ser, Thr, and Pro include backbone atoms; the others contain side-chain atoms only.

3 Packing of residues in protein interiors

F. M. Richards, C. Chothia, and their co-workers first emphasized the importance of the packing of residues in protein interiors to the definition and stabilization of native protein conformations. These ideas imposed a structure on W. J. Kauzmann's original 'oil drop' model of hydrophobic interactions (developed *before* atomic structures of proteins were known) which, although it correctly predicted the tendency of non-polar side-chains to form the protein interior, did not go so far as to indicate the geometric nature of the interactions. The observation that proteins have packing densities as high as those of amino acid crystals (35), typifies the results of the Yale school that led to the picture of native protein structures as solved jigsaw puzzles. (This idea must be carefully distinguished from the solving-the-jigsaw-puzzle model of the protein folding process.) Chothia and his co-workers have carried out a comprehensive series of studies on the geometric specificity of interactions between secondary structural elements in proteins—helix–helix, sheet–sheet, and helix–sheet—that determine tertiary structures (36).

This discussion has led us to the subject of tertiary structural interactions, and how to represent them. We have already pointed out that wire or ball-and-stick models are inadequate because they do not reproduce the space-filling properties of atoms. Shaded-sphere drawings are inadequate because one can see only one layer of atoms at a time, although it is possible to create serial sections by a succession of cheese-wire slices. Indeed, given the appropriate hardware to permit a 'hither' clipping plane to be moved interactively in real time, such pictures can be very effective.

For the time being, though, a useful alternative is to return to line graphics and draw slices through the Van der Waals envelopes of the atoms of a protein, as shown in *Figure 3.2*. We shall first discuss how to draw a single such section, and then describe some variations on the theme, including the simultaneous display of several successive serial sections, with and without hidden-line removal, and in combination with wire models.

Table 3.1 gives a list of suitable atomic radii.

Table 3.1. Van der Waals radii of atoms, in Å, useful for drawing pictures of proteins.

These radii are chosen to include the effects of hydrogen atoms, not explicitly present in protein structures determined by X-ray crystallography; for example, the radius of a tetragonal carbon (C—) is really the effective radius of a methyl group. The letter A represents the atoms in the amide group of an asn or gln residue, in case it is impossible to identify which is the N and which is the O. N3 = tertiary nitrogen; N4 = quaternary nitrogen, (C. Chothia and T. F. Koetzle, personal communication; compare ref. 37.)

H	C—	C=	N3		N4	A	O	F
1.00	1.87	1.76	1.65		1.50	1.52	1.40	1.35
				P			S	Cl
				1.70			1.85	1.80
Ca		Fe	Cu					Br
1.70		1.70	0.65					1.95
								I
								2.15

3.1 Van der Waals sections

Suppose we are given a set of atomic coordinates $(x_i, y_i, z_i, i = 1, \ldots n)$ and a Van der Waals radius r_i for each atom. We want to view the molecule from the positive z-axis towards the origin, cut through it by the plane $z = c$, and draw the Van der Waals envelope of the atoms that this plane intersects.

Three problems must be solved:

1. Identifying the atoms that appear on the section, and creating the circles that

represent the intersection of the cutting plane with the Van der Waals spheres around the atom.

Solution: An atom will be visible if and only if $|z_i - c| \leqslant r_i$. Visible atoms produce circles centred at (x_i, y_i, c) of radius $\sqrt{[r_i^2 - (z_i - c)^2]}$.

2. Identifying pairs of intersecting circles, and deleting the arcs of each that are in the interior of the other.

   Solution: To simplify the notation, let us work in two dimensions and suppose that we have a set of circles centred at (x_i, y_i) of radius ρ_i. For any two circles i and j, the centre-to-centre distance $d_{ij} = \sqrt{[(x_i - x_j)^2 + (y_i - y_j)^2]}$. Two circles will intersect if $d_{ij}^2 < (\rho_i + \rho_j)^2$.

   There is a technique for screening the set of circles for close pairs, so as not to have to check every pair of circles for intersection. We shall describe this method here for the two-dimensional case, but it is also very effective for extracting pairs of neighbouring atoms in three dimensions (38).

   Suppose that we overlay a square array similar to an extended chessboard on the section we have drawn, such that the edge of a square is 2max (ρ_i). For each square, make a list of the atoms the centres of which lie in that square. Then for any atom, we need only check for intersection those atoms the centres of which lie in the same square, or in the eight adjacent squares (if the atom were a chesspiece, namely a king, these comprise the square it occupies and those to which it could move). Any pairs of atoms that are not in the same or adjacent squares must have their centres further apart than 2 max (ρ_i) and cannot intersect. (The development of efficient algorithms for this type of problem, obviously of great importance in computational molecular biology, is currently a flourishing field of computer science called Computational Geometry; see, e.g., ref. 39).

   After identifying an intersecting pair of circles, we must compute the arcs that are to be deleted. This is easily done using the law of cosines, as shown in *Figure 3.5*.

3. Drawing the surviving arcs.

   Solution: The following procedure may be used to assemble the arcs into visible and deleted segments: For each circle, each individual pair overlap calculation produces two angles, α and β, for which the segment from α to β is to be *deleted*. Ensure that for each arc $\alpha \leqslant \beta$. Merge all the α's and β's into a single array, and sort the values, remembering which values are α's and which are β's. This divides the circle into a number of intervals, which might not overlap: $\alpha_1 < \beta_1 < \alpha_2 < \beta_2 \ldots$ or which might overlap: $\alpha_1 < \alpha_2 < \beta_2 < \beta_1 \ldots$

   The indices refer to the order of the angles *after* sorting. The first point must be an α, signifying the beginning of an invisible segment. If there are no overlaps, draw arcs from $\beta_1 - \alpha_2, \beta_2 - \alpha_3, \ldots \beta_n - \alpha_1$. If α_1 is not followed immediately by β_1, mark for deletion every entry between α_1 and β_1. Then find α_2 and if it is not followed immediately by β_2, mark for deletion all angles

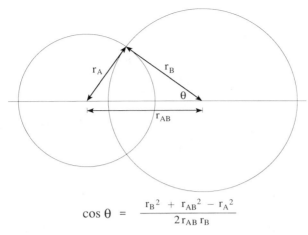

$$\cos \theta = \frac{r_B{}^2 + r_{AB}{}^2 - r_A{}^2}{2 r_{AB} r_B}$$

Figure 3.5 Deletion of overlapping arcs of two circles.

between α_2 and β_2. Continue in this manner until the end of the list, then delete all angles marked during the process. This produces a list of non-overlapping arcs, which can be drawn as described.

3.2 Applications

3.2.1 Density of atomic packing in protein interiors

Figure 3.6 shows a succession of serial sections through a protein, illustrating the solved jigsaw puzzle theme: there is relatively little free space. A systematic survey of pictures such as these or their equivalents reveals that holes large enough to contain a water molecule are rare in protein structures (40).

3.2.2 The geometry of helix–helix packing

Chothia and his colleagues have shown that the relative orientations of packed α-helices in protein interiors are determined by the topography of the helix surfaces. Side-chains adjacent on the helix surface tend to form ridges, with grooves between them. The helices pack with the ridges of one helix fitting into the grooves of the other. There are only a few ways to form the ridges; for example, they most frequently arise from residues four apart in the amino acid sequence, creating a ridge nearly parallel to the helix axis. Less frequently, ridges are formed from residues three positions apart, or adjacent in the sequence. The relative prominence of these potential ridges depends on the amino acid sequence and the side-chain conformation. *Figure 3.7* shows a well-developed ridge formed from residues four positions apart in the sequence.

Aligning the ridges of one helix with the grooves of another fixes, to a first approximation, the angle between the helix axes: indeed, Chothia and his co-

Figure 3.6 Serial sections through flavodoxin [3FXN], to show that protein interiors are densely packed. Flavodoxin contains a five-stranded β-sheet with two helices packed against each side, and contains the prosthetic group FMN [see part (a) and also *Figure 4.21a*]. Part (a) shows the orientation of the structure: the view is perpendicular to the sheet, with the prosthetic group at the top. Parts (b)–(g) show sets of serial sections through Van der Waals envelopes of the atoms. Each part contains three successive sections cut 1 Å apart. The first slice on each section in parts (c)–(g) is cut 1 Å beyond the last section of the previous part. The prosthetic group is shown by broken contours; the protein by solid contours. Labels are placed at the positions of Cα atoms.

It takes a little practice to recognize secondary structures in such diagrams (an overlay as shown in part (h) can be quite helpful). Note the following: In part (b) there is a vertical helix at the left (residues 121–137). Compare *Plate 1*. Note that the helix is approximately parallel to the strands of the sheet against which it packs. Parts (d) and (e) contain strands of β-sheet. In part (e) the slices pass through the backbone of regions (from the left) 111–116, 83–88, 49–54, and 4–8. (A keen eye will detect the pleating of the sheets in the positions of the labels.) Compare part (d) in which the slices pass through the alternate side-chains protruding above the sheet—seen most clearly for residues 48–50–52. In parts (f) and (g) the ridge formed by the side-chains of residues 65–69–73 on the surface of a helix can be seen. Part (g) shows the flavin ring of the prosthetic group seen end-on.

In case these diagrams remain obscure, part (h) contains the same slices as part (f), with a wire model overlaid.

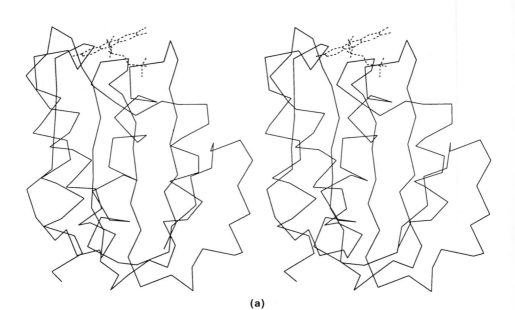

(a)

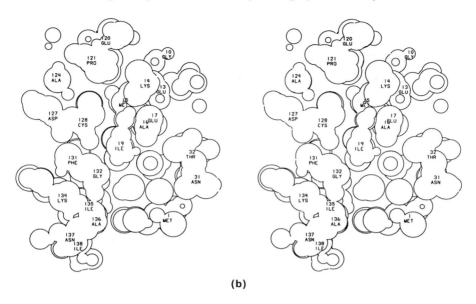

(b)

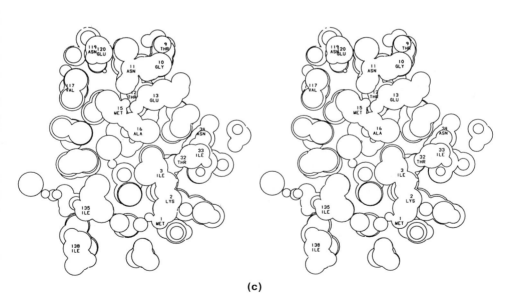

(c)

Parts d, e are overleaf

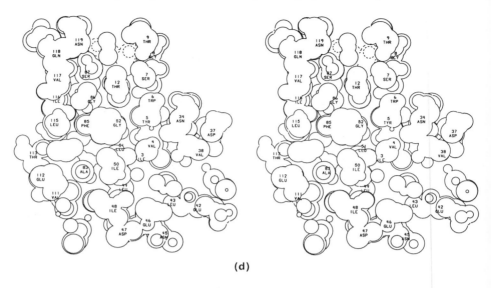

(d)

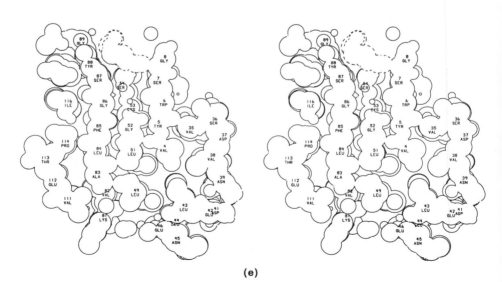

(e)

54

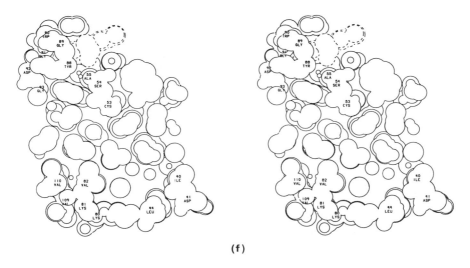

(f)

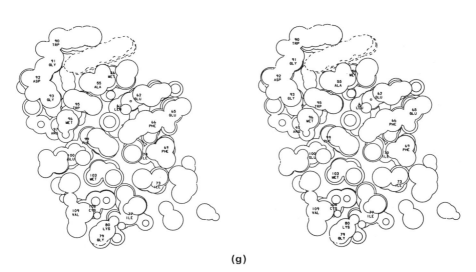

(g)

Part h is overleaf

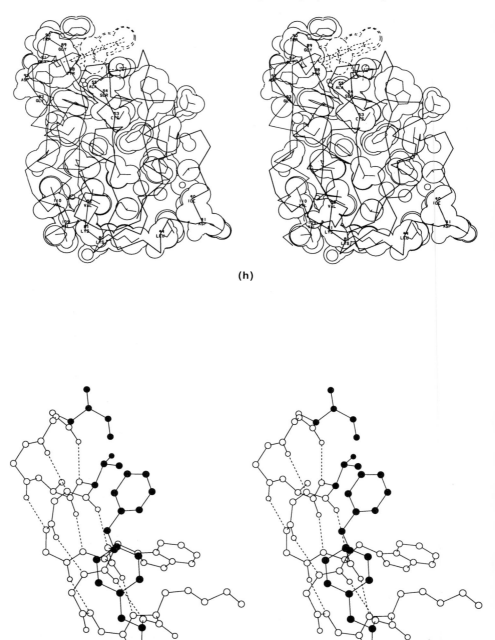

(h)

Figure 3.7 A ridge along the surface of an *α*-helix from carboxypeptidase a [5CPA].

workers observed a peaked distribution of inter-helix-axis angles that correlated with the nature of the packing and agreed with the theoretical predictions (41, 42). This topic will be developed further in the next chapter.

In *Figure 3.8* we show the use of superposed serial sections and skeletal models to show the packing at a typical helix interface. It is often effective to show three serial sections spaced 1 Å apart, together with a wire model and labels. Note that the two helices can be distinguished by solid vs. broken contours, and that hidden line removal can, if desired, be applied to the Van der Waals sections but not to the wire model. In this way one can 'eat one's cake and have it too': taking advantage of the enhancement of perception of spatial relationships derivable from hidden line removal without entirely sacrificing transparency, as raster shaded sphere pictures typically do. It can be verified in the wink of an eye that stereo is useful.

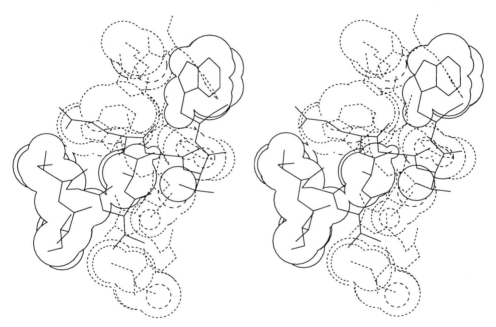

Figure 3.8 The packing in the interface between helices in carboxypeptidase a [5CPA].

4 Dot-surface drawings

There is another way to combine the space-filling features of Van der Waals surfaces with transparency allowing the mixing of skeletal models. This is to 'spatter-paint' dots on the surface to be shown. *Figure 3.3* and *Plates* 2 and *3* showed examples of this technique, but, as we have already mentioned, it is most effective when displayed on an interactive graphics device with a colour screen.

Used on such a system, dot-surface pictures have had many successes in the problem of 'docking' ligands to proteins and exploring the goodness of fit in interfaces.

Technically, the problem of distributing N points as evenly as possible on a sphere is a difficult one. Of course, it is well-known that there are only 5 regular polyhedra, so that it is not possible to distribute N points on a sphere with full symmetry in their distribution unless N is the number of vertices of one of these. The largest of these, the icosahedron, has 12 vertices, but for the pictures we wish to draw, several hundred points per sphere are desirable. A pragmatic solution is to choose a nearest-neighbour angular displacement, and distribute points on circles of constant latitude keeping the angular displacement as close to constant as possible.

Mathematicians have taken an interest in this problem, as it is important in numerical integration on the surface of a sphere [the calculation of accessible surface area by the method of Shrake and Rupley (43) is an example of this], and there is a sizeable literature on the subject. Sloane and Conway (44) have published tables showing optimal distributions for various values of N. (Of course, the problem is also of interest to virologists analysing the packing of molecules in coats of spherical viruses.)

After distributing dots on each of the atoms, it is important to remove dots from each atom that lies within the spheres corresponding to other atoms, to produce a simple non-selfinterpenetrating surface. This can be accomplished by methods that are a fairly direct generalization to three dimensions of the techniques described in Section 3.1 for removing arcs from intersecting circles.

Table 3.2 compares the different ways of representing atoms as spherical surfaces.

Table 3.2 Representation of atoms by spherical surfaces

Type of imaging device	Appearance of atom	How different atom types are distinguished	How depth relationships are represented
Plotter	discs (spatter-surfaces, wire networks)	size, shading, colour	stereo, hidden-line removal
Static raster display	shaded sphere (spatter surface)	size, colour	stereo, hidden-surface removal, depth cueing
Interactive calligraphic display	spatter surface	size, colour	kinetic depth effect, stereo, depth cueing
New powerful specialized graphics workstation	all of the above	all of the above	all of the above

5 Conclusion

In this chapter we have tried to emphasize the contrast between the qualities of skeletal models produced on vector graphics devices with those of shaded sphere pictures produced on passive raster devices. These two representations can be considered as extremes, and, probably because they are extremes, they have turned out not to be the most useful. We have already seen in this chapter the value of having the versatility to create different representations, and to mix them in the same picture. In the next chapter we will develop this theme further, introducing many new representations, and discussing their particular virtues and how they can most effectively be combined.

4

Pattern and form in protein structure

1 Introduction

It will have become clear in previous chapters that the traditional representations are not adequate to solve the problems of illustrating and analysing complex structures. What is necessary is a simplified representation that still retains the important features of the molecule: a schematic diagram, or cartoon. Such simplifications were devised by A. Rossmann, and by A. Liljas and B. Furugren. Helices can be depicted as cylinders, and β-sheets by thick arrows (see *Figure 4.1*). Following their lead, many other people have drawn such pictures by hand, those of J. Richardson (45) being most widely known.

This does not mean, of course, that one must discard the more detailed representations. When K. Hardman and the author designed a computer program that included facilities for drawing schematic diagrams, it was our idea to compile into one program the facility to produce as many different representations as we could (46). This gave the user the ability to mix different representations within a single picture, which proved extremely valuable. Not only did a variety of combinations of representations turn out to be more useful than we were insightful enough to foresee (see, for example, *Figure 3.8*), but it became possible to 'fine-tune' the level of detail with which different parts of a molecule were portrayed. For example, a binding site might be shown in full atomic detail, while the remainder of the molecule remained simplified. (See *Figure 1.3*.)

Other obvious advantages of generating pictures by computer rather than by hand are that it is possible to vary the orientation easily, that it is possible to produce stereo pairs (not impossible but extremely difficult to do by hand) and that the result can be displayed on an interactive graphics device. A number of such programs are now available (see for example, ref. 47).

In this chapter we shall explore both different representations (and combinations of representations) and different proteins.

Because an important application of such pictures is the comparison and classification of protein structures, we shall develop the background of these problems first.

61

Figure 4.1 The schematic representation of (a) *α*-helices as cylinders and (b, c) strands of *β*-sheet by large thick arrows. Note the 'twist' of the sheet. Compare *Figure 2.2*.

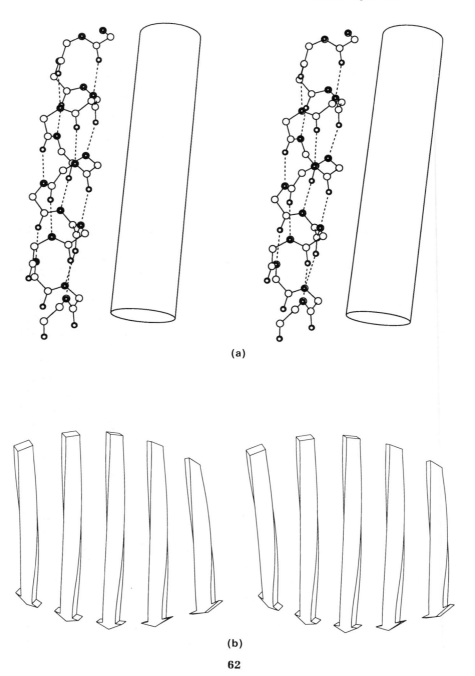

(a)

(b)

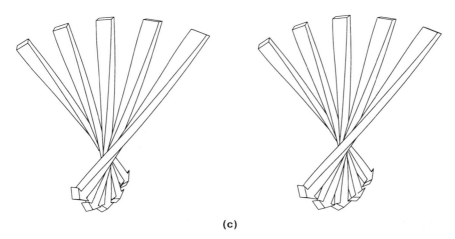

(c)

2 A problem: defining the extents of helices and sheets

If helices and sheets are to be represented by a schematic 'icon' computed from the coordinates, it is necessary to decide where these regions of secondary structure begin and end.

It is a fact of life that helices and sheets in globular proteins can be irregular, especially at their ends. Helices can bend in the middle, to the extent of losing a hydrogen bond or two (often but not exclusively at a proline), and frequently either unravel or tighten up in their last turn. β-sheets may be interrupted by β-bulges (48) (*Figure 4.2*), and can also show irregularities at ends of strands.

Of course, these effects make any definition of the limits of the regions of secondary structures somewhat imprecise. *Indeed, for different purposes different definitions may be appropriate and necessary.* For example, if one wants to determine all the pairs of residues in contact at a helix–helix interface, the most generous definition of the helical regions should be used. But to measure the inter-axial distance and angle between two helices, the regions should be restricted to those residues that form a regular helix, omitting frayed ends.

Protein Data Bank files contain secondary structural assignments reported by individual crystallographers. However, crystallographers do not all agree on the criteria for deciding where regions of secondary structure begin and end. In consequence, there have been attempts to write computer programs to identify regions of secondary structures consistently and objectively. The first of these was by M. Levitt and J. Greer (49), and there is a more recent one by W. Kabsch and C. Sander (50). These programs do have the merit that they apply the same criteria, in a consistent way, to all proteins (indeed, it is essential to have an objective way to assign secondary structure, for tests of secondary-structure prediction algorithms). What is unfortunate is that many people use these secondary structure assignments unquestioningly; perhaps the greatest damage

63

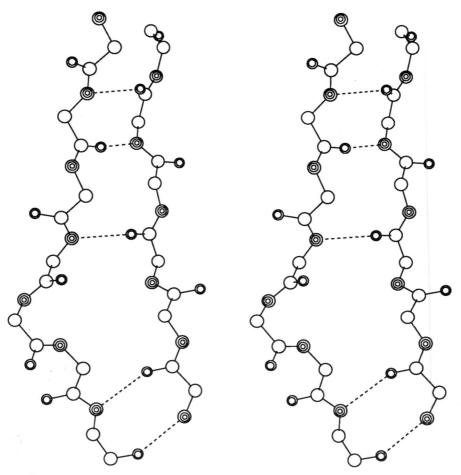

Figure 4.2 A β-bulge in a sheet in immunoglobulin fragment V_L J539 [1FBJ].

done is to create an impression (for which Levitt, Greer, Kabsch, and Sander cannot be blamed) that there is **a right answer**. Provided that the danger is recognized, such programs can be useful.

The reader will have noticed that in the last paragraph I have expressed dissatisfaction with *both* crystallographers' assignments of secondary structures and the use of programs. What is left? What remains is the inspection of pictures of regions in proteins, together with lists of hydrogen bonding patterns and conformational angles, and a clear understanding of what decision about the boundaries of elements of secondary structure is appropriate for the application at hand.

3 Classification of protein structures

When Linnaeus created his classification of living things, he was making a catalogue of similarities among the objects in the corpus of material he was studying. Only later did it emerge that the hierarchy of relationships that Linnaeus observed was indeed induced by evolutionary processes and reflected a closeness of biological kinship.

With protein structures, we can observe similarities in topology or even in structural details. But only in *some* cases can we infer genuine biological relationships in the sense of descent from a common ancestor. That is, there are many well-characterized protein 'families': for example, we know the structures of globins from 14 species, and the sequences of several hundred. Analysis of the sequences and structures shows that one can construct evolutionary trees on the basis of molecular data, from related molecules in different species, and these have been shown in most cases to be equivalent to the evolutionary trees constructed from classical comparative anatomy (for a review see ref. 51).

Within protein families, classification parallels ordinary taxonomy. It is when we compare members of *different* protein families that we can perform only a rather gross classification, and one that does not lend itself to the formation of a hierarchy that can be extended to many levels by drawing less- and less-detailed distinctions. This situation may reflect a history in which proteins of different families were *not* descended from a common ancestor, or it may reflect our incomplete understanding of the processes of structural change.

It may be instructive to pursue the analogy with biological classification a bit further. Comparing human haemoglobin with dog haemoglobin is like comparing the skeleton of a human arm with the skeleton of a dog's foreleg; it is well-known that in the case of the skeletons there is qualitatively a bone-for-bone correspondence, and a topologically similar arrangement, but that quantitatively the sizes and shapes of the corresponding bones differ. But comparing a human haemoglobin with human (or dog) chymotrypsin is like comparing a lung with a stomach. In this case, only an understanding of the evolutionary history of these organs can assess the nature of the relationship. Superficial similarities in structure and function may not reflect a truly close relationship (as in the case of the eye of a human and the eye of an insect) and divergent evolution may conceal a homology (for example, bones of the human ear are descended from bones which in fish formed part of the jaw).

3.1. Measures of similarity of protein sequences and structure

In devising measures of similarity and difference between two proteins, it is sometimes clearer to see how to proceed in comparing sequences, than in comparing structures. If the amino acid sequences of two proteins can be aligned, then we can either count the number of identical residues, or use a more subtle metric based on a similarity index between amino acids. Such an index would take the form of a 20×20 matrix, M, such that each entry corresponds to a pair of amino acids and M_{ij} gives a measure of the similarity between any pair of amino acids. The similarity between two sequences is then the sum of values for each pair of aligned amino acids, taken from the matrix, plus a correction to account for the gaps found in the sequences at sites of insertions or deletions of amino acids. Indeed, it is by maximizing such a similarity score that the optimal alignment of two sequences is conventionally calculated (52, 53).

In three dimensions the problem can be more complex. If two protein structures are very closely related, and we can align the residues, it is then easy to superpose the corresponding residues, and to measure and analyse the nature of the deviations in position of corresponding atoms. (A quantitative analysis of structural difference and structural change is the subject of the next chapter.) Structure tends to change more conservatively than sequence, and it is not uncommon to be able to recognize a relationship between proteins from their structures, when no evidence of homology appears in the sequences. The two domains of rhodanese are a classic example.

However, comparing proteins from within the same family is a special case. In studying and attempting to classify the corpus of known proteins as a whole, one is dealing in most cases with molecules for which no sequence alignment is possible; that is, for which no explicit residue–residue correspondence can be established. Nevertheless, one still observes a recurrence of structural themes among apparently unrelated proteins, and it is of great interest to classify them on the basis of secondary and tertiary structures, and on topology or 'fold'. The implications of these structural similarities in terms of evolutionary relationships are in many cases still obscure.

3.2 Classification of protein topologies

The most useful classification of families of protein structures (that is, a classification general enough to encompass even unrelated proteins) is based initially on work of Levitt and Chothia (54). Their classification is grounded on the general properties of secondary and tertiary structure in proteins. Protein structures (more precisely, domains within protein structures) can be classified according to the categories in *Table 4.1*.

One use to which a hierarchical classification of proteins could be put would be in assessing the quality of our ability to predict protein structures from amino acid sequences. Ultimately, we should like a fully detailed prediction of the atomic structure. This we cannot now do. What can we do? Programs to predict secondary structures can now distinguish helix, sheet and turn residues with over 60% accuracy. This can do a fairly good job of distinguishing the basic classes of protein families: α, β, $\alpha + \beta$, and α/β (55). Other types of programs seek to distinguish one protein family, e.g. the globins, from other families, based on patterns of conservation of residues in the aligned amino acid sequences of proteins of that family (see refs 56–59 for the specific case of the globins, and refs 60 and 61 for general reviews).

Table 4.1. Classification of protein topologies

Property	Class	Characteristic	Examples
Secondary structure content			
	1. α-helical	secondary structure almost exclusively α-helical	myoglobin, cytochrome c, citrate synthase
	2. β-sheet	secondary structure almost exclusively β-sheet	chymotrypsin, immunoglobulin domain
	3. α+β and α/β	secondary structure contains both α-helix and β-sheet	papain, alcohol dehydrogenase, triose phosphate isomerase
Tertiary structure			
	2.1 parallel β-sheet	double β-sheet sandwich, strands roughly parallel	immunoglobin domain
	2.2 orthogonal β-sheet	double β-sheet sandwich, strands of different sheets roughly perpendicular	chymotrypsin domains
	2.3 β-sheet (other)	other types of spatial relationships between sheets	neuraminidase, interleukin-1β
	3.1 α+β	α-helices and strands of β-sheet separated in different parts of molecule. Absence of β–α–β supersecondary structure	papain, staphylococcal nuclease
	3.2 α/β	helices and sheet assembled from β–α–β units—strands of sheet parallel	alcohol dehydrogenase, triose phosphate isomerase
	3.2.1 α/β-linear	line through centres of strands of sheet roughly linear	alcohol dehydrogenase
	3.2.2 α/β-closed	line through centres of strands of sheet roughly circular	triose phosphate isomerase

Topology—at this point it becomes difficult to tabulate a set of mutually-exclusive classes. The distinction is that two parallel β-sheet proteins—for example, plastocyanin and the immunoglobulin domains—may have similar constellations of elements of secondary structure, but differ in their connectivity: that is, in the order of the strands along the sequence. This is enough to suggest that they are not related by evolution, as there does not appear to be a possible continuous path from one topology to the other, that could be followed by molecules produced by a sequence of point mutations.

However, the most powerful techniques for making detailed predictions of protein structures work only *within* a single protein family. This is currently the main hope of giving an intelligent answer to the people who approach computational molecular biologists with the question: 'Here is a newly-determined amino acid sequence; what can you tell me about the structure and function of the protein?'

We begin the discussion of this question in the next section.

3.3. 'Here is a new sequence . . .'—Part I

Applications of the principles outlined in previous sections arise fairly frequently nowadays, when the nucleotide sequences of genes coding for unknown proteins are determined. There is a fairly well worked-out procedure for responding to the question: 'What can you say about the structure or function of this protein?'

First, one carries out a 'screen' of the new sequence against a database of known sequences. This answers the following questions:

1. Is there any protein of known structure that has sufficient similarity to the sequence of the unknown protein to suggest a familial relationship?

2. If not, what sequence of any known protein is most similar to the sequence of the unknown protein?

In the most favourable case, the new protein will be recognizable as a member of a family of proteins of known structure. If, in an optimal alignment, over 25% of the residues are identical, it is fairly safe to conclude that the two proteins have the same fold. Note that for distantly-related proteins sequence alignment will not always provide a reliable answer: the sequences of two related proteins may have less than 20% residue identity in an optimal alignment, and the sequences of definitely unrelated proteins may have as much as 20% residue identity in an optimal alignment (*Figure 4.3*). Several authors specializing in sequence analysis have proposed criteria for accepting a tenuous relationship as genuine, but borderline cases remain doubtful.

In a favourable case, when one can identify a relationship to a protein of known

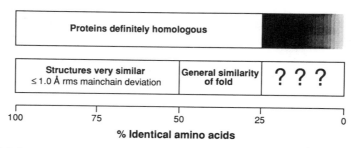

Figure 4.3 Sequence similarity and evolutionary relationships. Adequate similarity between sequences can prove evolutionary relationship, but lack of sequence similarity cannot disprove it.

structure, it is possible to suggest that the new protein shares a common structure with its relative and to assign its general fold. It is possible to go on to give a quantitative assessment of the structural similarity, based on how far the amino acid sequences have diverged. This will be discussed in the next chapter.

Suppose, next, that the new protein sequence is related to a known protein of unknown structure. If the known protein has a known function, the unknown protein will be expected to have a similar or at least a related activity. Exceptions do exist, however. For instance, the serine protease inhibitor, α_1-antitrypsin, is homologous to the storage protein ovalbumin.

If database screening methods based on overall sequence similarity do not identify a relative, there is one more string to our bow. In a number of cases, the active site of a protein can be recognized by a specific 'fingerprint' or 'template', a constellation of a fairly small set of residues that are unique to a family of proteins. An example is the sequence

$$G*G**G$$

(where G = glycine and * = any amino acid) which seems to define a binding site for GTP. Hodgeman (60, 61), and A. Bairoch have tabulated sets of known templates. This can be quite a powerful technique, although we do not yet have templates for all protein families.

The development of this background has been necessary so that the reader will understand what kinds of pictures ought to be drawn. We can now proceed to discuss how to draw them.

4 Schematic representations

4.1 Summary

We begin by reminding the reader of the fundamental dilemma in designing an illustration of a protein structure: Many minute details are of the utmost importance, but a picture containing too many details is unintelligible. In this section we describe some of the simplified representations of protein structures that have been used to show the overall folding pattern or topology. Later we shall explore how to integrate these schematic representations with detailed representations of selected portions of the molecule, to produce pictures showing both the 'gestalt', and the relation to the gestalt of selected structural details. Thus the reader is urged to regard each of the representations described as a potential ingredient in a carefully-chosen mixture. Designing an effective illustration of a single protein structure, a comparison of two structures, or a conformational change is often quite a tricky task. The standard advice to naval recruits: 'If it moves, salute it; if it doesn't move, paint it' (62), is extremely useful but does not solve the problem completely.

4.2 Tracing the chain

The simplest representation of the fold of a polypeptide chain is a set of line segments that link successive α-carbons. Stereo is almost essential; depth cueing (difficult but not impossible to achieve in a drawing on paper, but generally available on interactive graphics devices) is very helpful in restoring the third dimension to these drawings. Small arrowheads placed with discretion along the chain help the eye to keep track of the chain direction. See *Figure 4.4*.

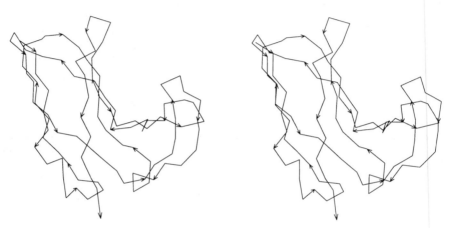

Figure 4.4 Representation of actinoxanthin by a tracing of the chain with line segments connecting successive Cα's [1ACX].

A generalization developed many years ago by McLachlan and Blow is to represent the main chain of each residue as a small polygon in the plane of the peptide group, linked together to form a kinked ribbon. In addition to carrying information about the tilts of the peptides, this representation permits hidden-line removal to assist in depth perception. (See *Figure 4.5*.) Even for quite large structures this representation can provide a clear exposition of the fold. (See *Figure 4.5d*.) The combination of the 'polygonal ribbon representation' with a simple chain trace can be a useful way to distinguish strands from loops in β-sheet proteins. (See *Figure 4.5e*.)

Each of these representations can be generalized by replacing the segmented representation by a smooth curve or ribbon through the same points. These bear roughly the same relationship to the segmented representations as cursive writing does to printing. By reducing still further the amount of detail, these pictures provide no complication or distraction from the overall folding pattern. (See *Figure 4.6*.) The chain can be represented either as a thin or thick line or as a ribbon, as in Appendix 3.

Figure 4.5 'Ribbon' diagrams.

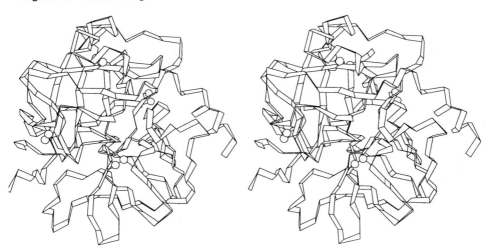

(a) *a*-chymotrypsin [2CHA]. This molecule, like other serine proteases, contains two domains believed to have originated by gene duplication and divergence. Each domain contains two β-sheets, packed face to face, with the strands of the different sheets running in approximately orthogonal directions. One such pattern of 'crossed' strands can be seen in the upper left of the molecule in this orientation, with the strands of the sheet in front running approximately horizontally and those in the sheet in the back running approximately vertically.

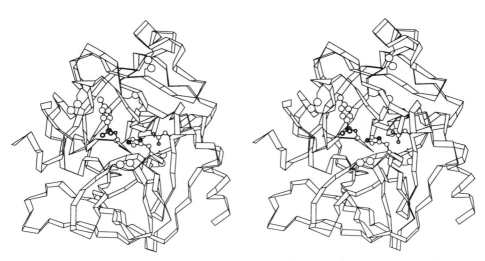

(b) Another view of *a*-chymotrypsin [2CHA], showing the catalytic triad (from right to left) Asp 102, His 57 and Ser 195. The serine is covalently linked to a toluenesulphonyl group, the ring of which occupies the 'specificity pocket' of the enzyme.

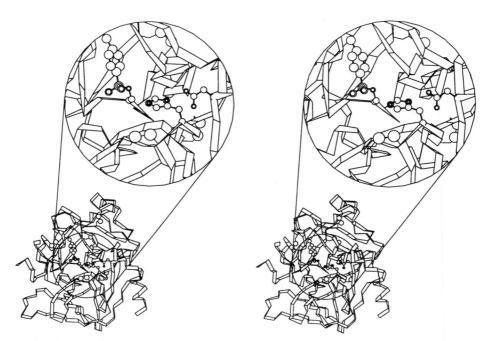

(c) The same view of *α*-chymotrypsin [2CHA] as part (b), with the region around the catalytic triad and inhibitor 'blown up' for clarity, but still showing the relationship of these residues to the overall structure.

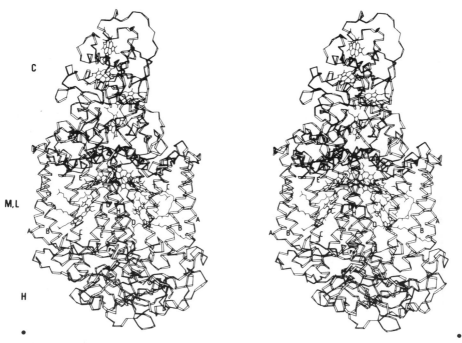

(d) The reaction centre of *Rhodopseudomonas viridis* [1PRC]. (From ref. 63.)

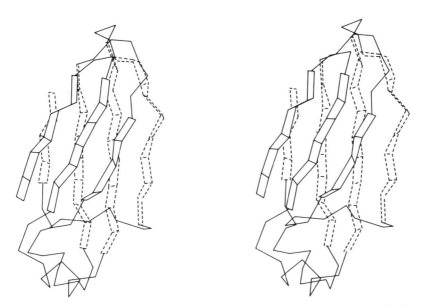

(e) Immunoglobulin constant domain. This is also a double β-sheet structure, but in this case the strands run in approximately parallel directions. The strands on the sheet in front are drawn as a ribbon with solid lines; those of the sheet in the back are drawn as a ribbon of broken lines, and the links between them are drawn as a simple chain trace.

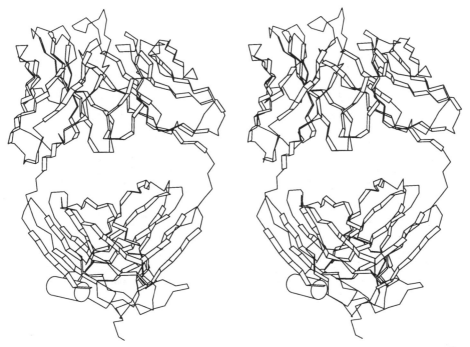

(f) An immunoglobulin Fab fragment, showing four domains, each of which is generally similar to the C_L domain shown in part (c) [1FB4].

73

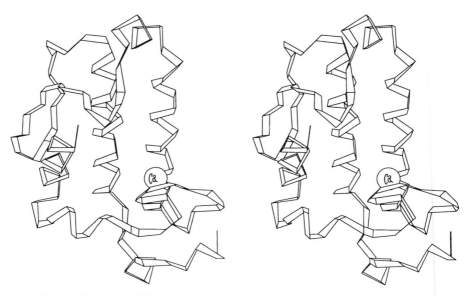

(g) Phospholipase A$_2$ [1BP2]. In this and similar drawings, an attempt has been made to enhance the appearance of three-dimensionality of the isolated atom or ion by printing its chemical symbol on the surface of a sphere.

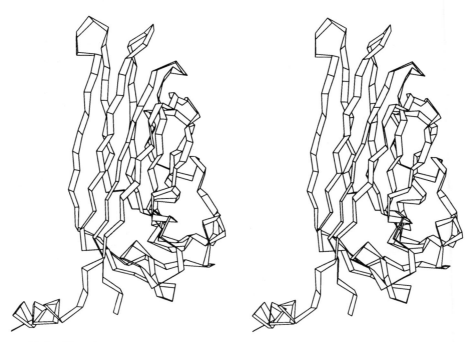

(h) Satellite tobacco necrosis virus coat protein [2STV].

A related representation developed for colour raster devices shows the chain tracing as a jointed tube (see *Plate 4*).

Still another representation that can be useful shows the chain as a 'string of pearls', which includes sequence information in a concise way. See *Figure 4.7*, and, for a specialized application, Section 5.4.3 of Chapter 4.

4.3 The representation of helices by cylinders

The use of a cylinder as an 'icon' for a helix is one of the oldest and most common of schematic representations. The axis of the cylinder coincides with the axis of the helix, and usually the cylinder extends to the positions (projected on to the cylinder axis) of the first and last α-carbon of the helix.

The radius of the helix can be chosen as the radius of the α-carbon relative to the helix axis (in cylindrical polar coordinates). This is 2.29 Å for an α-helix and 1.91 Å for a 3_{10} helix. Again, these radii provide compatibility with other representations if, for example, one wants to add selected side-chains to the helix. However, in pictures of very large proteins, with hidden-line removal used so that the helices appear opaque, it may be necessary to use a smaller radius to avoid obscuring too much of the molecule.

Combinations of chain tracing and schematic representations of secondary structural elements are useful to show the general architectural pattern of a molecule. *Figure 4.8* shows the rattlesnake venom phospholipase dimer.

4.4 The representation of strands of sheet by large arrows

This is the second of the early 'icons' for elements of secondary structure. A strand of sheet is depicted by a large arrow—typically something like 2 Å wide and 0.5 Å thick, with an arrowhead at the C-terminus to indicate the chain direction.

The shaft of the arrow ought not to be a rectangular solid, but should be twisted to show the twist of the sheet. β-sheets in globular proteins are typically twisted rather than flat (see *Figure 2.2* (a, b)). This means that the strands, although individually relatively straight, do not all lie in the same plane. Computationally, showing the twist can be achieved by a smoothing procedure analogous to that applied to the simple chain-trace and polygonal ribbon representations.

As with ageing actors and actresses contemplating cosmetic surgery, there is considerable debate about how much smoothing should be done. Strands of β-sheets in proteins are not perfectly regular; oversmoothing will cause them to appear as if they were, and will produce a picture that may be aesthetically more satisfying but less accurate. A special problem with sheets is presented by the presence of β-bulges. These are residues—usually only one or two—that break the regular pattern of sheet hydrogen-bonding and the regular course of the chain. Not surprisingly, they can play havoc with non-robust smoothing procedures.

Figure 4.6 Smooth curves through the chain of (a) actinoxanthin [1ACX] (compare *Figure 4.4.*) (b, c) Phospholipase A₂ [1BP2] (compare *Figure 4.5g*). Note the effect on the helices of two different degrees of smoothing.

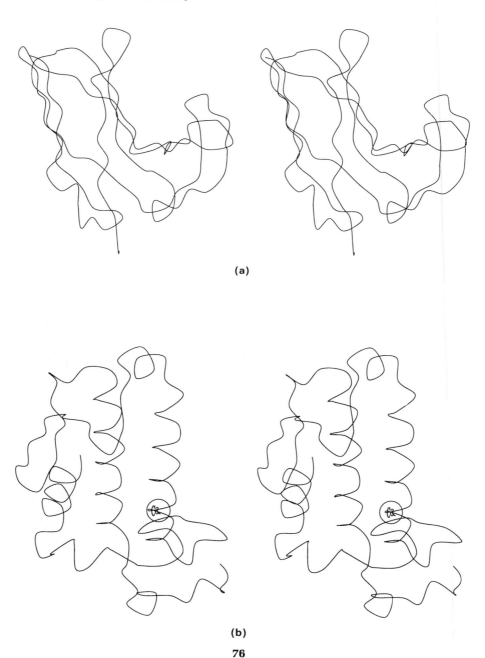

(a)

(b)

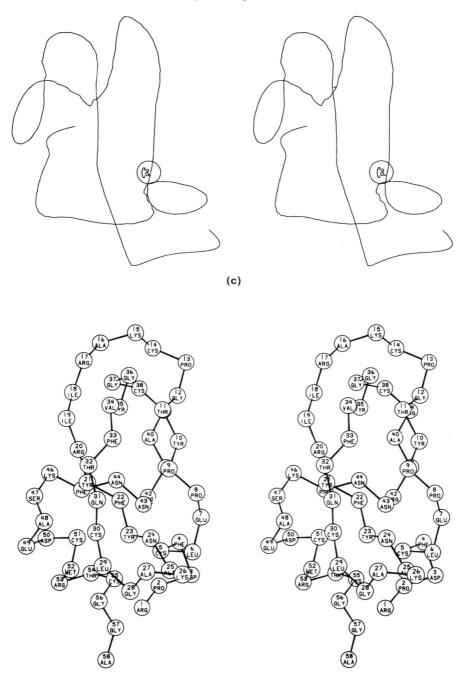

(c)

Figure 4.7 'String of pearls' representation of bovine pancreatic trypsin inhibitor [4PTI].

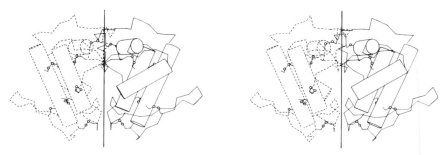

Figure 4.8 Rattlesnake venom phospholipase, a dimer [1 PP2]. One monomer is shown in solid lines; the other in broken lines. The vertical line in the centre is the dyad axis relating the two monomers. Each monomer contains seven disulphide bridges.

5 The varieties of protein structure

The purpose of this section is to show some applications of the representations we have discussed to the illustration of protein topologies. It is organized approximately according to the classification of Section 3.2, but it is not comprehensive.

5.1 Helical proteins

The α-helix had been predicted by Linus Pauling in 1950, long before the solution of the X-ray crystal structures of the first globular proteins by Kendrew and Perutz. These first structures, myoglobin and haemoglobin, were rich in helices. Of these, all but one are α-helices. The C-helix is a 3_{10} helix. Most proteins contain some helix; in a significant number the secondary structure is limited to α-helices. For example:

1. There is a class of small proteins that have the form of a four-helix bundle. (See *Figure 4.9*.)
2. The globins include monomeric proteins in many species including animals, plants and bacteria, and tetramers such as mammalian haemoglobins (*Plate 5*). It is well-known that the quaternary structure of haemoglobin is necessary for its allosteric properties. In each of these molecules the monomer binds a haem group flanked in most cases by two histidine residues.
3. The cytochrome c's are another family of haem proteins with exclusively helical secondary structures (*Figure 4.10*).
4. Even very large proteins can have a purely helical secondary structure, citrate synthase being an impressive example, illustrated in *Figure 5.7*.
5. Bacteriorhodopsin contains seven transmembrane helices (*Figure 4.11*).

Even from these examples it is clear that it is very difficult to circumscribe or even to classify the types of structures that collections of α-helices can create. In a very interesting development of polyhedral models, Murzin and Finkelstein (64)

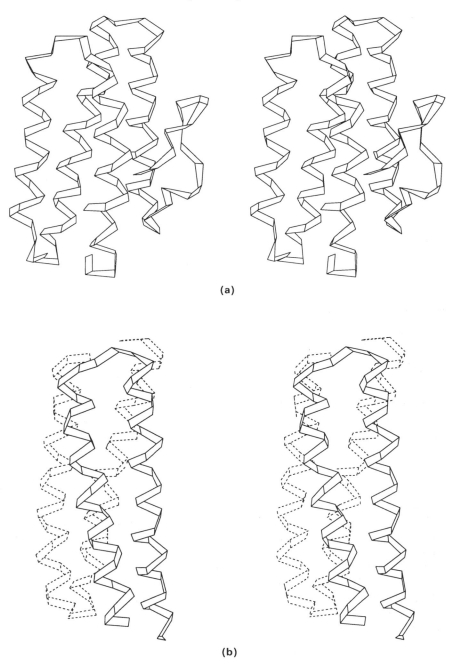

(a)

(b)

Figure 4.9 Four-helix bundles: (a) Haemerythrin [1HMQ], (b) ROP, 'Repressor of Primer', a dimeric four-helix bundle. (Courtesy of D. Banner.)

Figure 4.10 Members of the cytochrome c family. (a) Tuna cytochrome c [5CYT], (b) cytochrome c_{551} from *Pseudomonas aeruginosa* [451C], and (c) cytochrome c_2 from *Rhodospirillum rubrum* [3C2C].

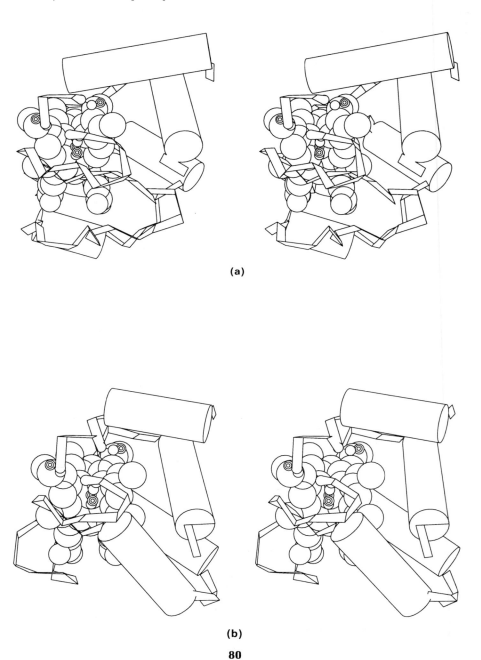

(a)

(b)

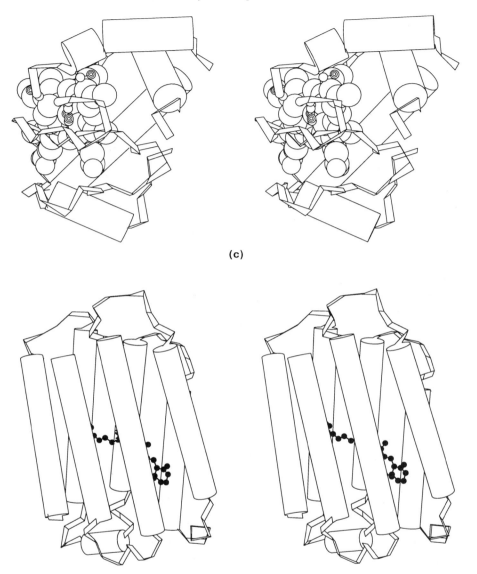

(c)

Figure 4.11 Bacteriorhodopsin. The helices traverse a membrane. This molecule has the function of a light-driven proton pump. (Courtesy of R. Henderson, J. M. Baldwin, T. A. Ceska, F. Zemlin, E. Beckmann, and K. H. Downing.)

have addressed the problem of the global assembly of helices in α-helical proteins. For individual pairs of interacting helices, certain regularities in the patterns of tertiary–structural interactions between pairs of helices have been observed. These will be discussed later in this chapter.

5.2 *β*-sheet proteins

Domains in which the secondary structure is almost exclusively β-sheet tend to contain two sheets packed face to face. There are two major classes: those in which the strands are almost parallel (like the fingers of Dürer's praying hands) and those in which the strands are almost perpendicular (like the fingers of two people shaking hands). Unlike the fingers of a hand, however, the strands of sheet can vary in direction (parallel or antiparallel) and in connectivity, giving great topological variety.

5.2.1 Parallel *β*-sheet proteins

Prealbumin is fairly typical of the small proteins that comprise this class (*Figure 4.12*). Each sheet contains 4 strands. It is thought that a greater number of strands would make it difficult to accommodate the natural twist of the sheet. In concanavalin A, for example, the large sheet is unusually flat (*Figures 2.2c and 4.13*).

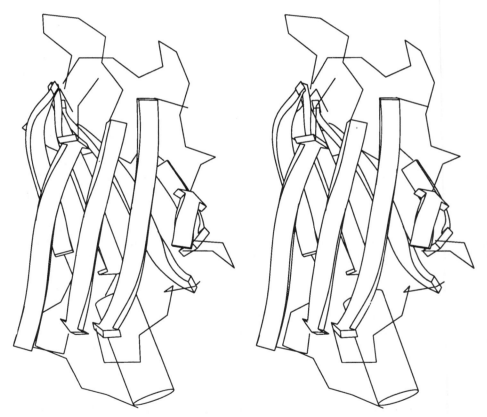

Figure 4.12 Prealbumin [2PAB].

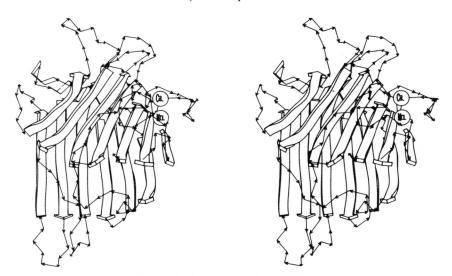

Figure 4.13 Concanavalin A [3CNA].

If we compare prealbumin with several other members of this class of structures, such as plastocyanin or the immunoglobulin domain, the similarities in the arrangements of secondary structural elements are striking (*Figure 4.14*). How can we tell whether these proteins are showing a common solution of a structural problem or an evolutionary relationship? Notice first of all that the directions of the strands are not all equivalent. Moreover, the connectivities of the strands are different.

Let us examine this in more detail. If we trace the chain along the sequence, we can examine the topological relationship between successive strands. For example, if we come up along one strand we might then find a 'hairpin' loop between two hydrogen-bonded strands on the same sheet, or we might find that the next strand encountered, as we proceed along the sequence, is on the opposite sheet. It is because it is difficult to envisage a continuous pathway whereby evolution could proceed to alter the topology of these connections, that it appears that these molecules are not related by evolution. They are merely showing a common solution of a general structural problem. (See ref. 65.)

5.2.2 Orthogonal *β*-sheet proteins

An alternative way of packing two *β*-sheets together is with the strands in the two sheets almost perpendicular. Each domain of the serine proteases shows this arrangement, as does retinol-binding protein (*Figure 4.15* and *Plate 6*).

5.2.3. Other *β*-sheet proteins

Ascorbate oxidase is a large *β*-sheet protein that contains parallel *β*-sheet domains (*Figure 4.16a*). Influenza neuraminidase contains a very unusual *β*-sheet

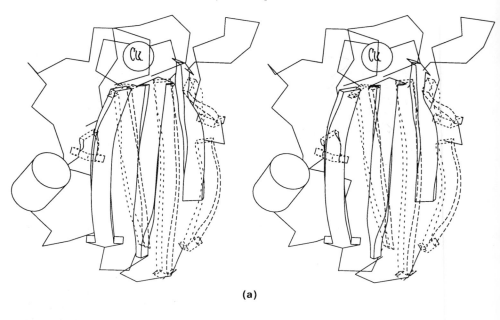

(a)

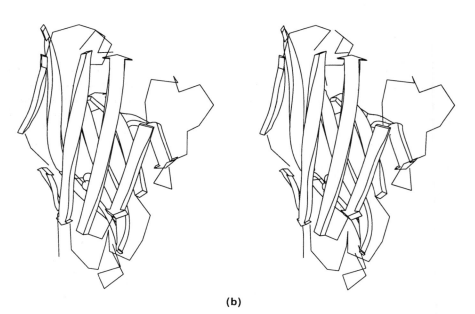

(b)

Figure 4.14 (a) Poplar leaf plastocyanin [1PCY]; (b) immunoglobulin domain (V_H KOL) [1FB4].

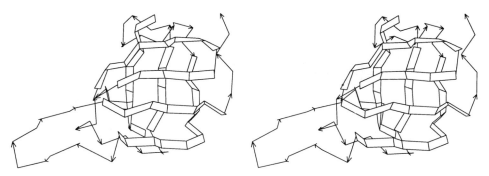

Figure 4.15 The N-terminal domain of γ-chymotrypsin [2GCH]. Compare *Figure 4.5a.*

'propellor' (*Figure 4.16b*). Interleukin-1β may be described as containing both a barrel and a propellor (*Figure 4.16c*).

5.3 α + β proteins

Many proteins contain both α-helices and β-sheets, but do not have the special structures created by alternating β–α–β patterns, considered in the next section.

In the sulphydryl protease actinidin and in staphylococcal nuclease, the strands of sheets and the helices tend to be segregated in different regions of space (*Figures 4.17* and *4.18*).

5.4 α/β proteins

5.4.1 The β–α–β unit

The 'supersecondary structure' consisting of a β–α–β unit, with the β-strands parallel and hydrogen-bonded, forms the basis of many enzymes, especially those that bind nucleotides or related molecules (*Figure 4.19*). The strands form a parallel β-sheet. In some cases, there is a linear β–α–β–α–β . . . arrangement, but in other cases the β-sheet closes on itself, the last strand hydrogen-bonded to the first.

5.4.2 Linear or open β–α–β proteins

Many proteins that bind nucleotides contain a domain made up of six β–α units, with a special topology:

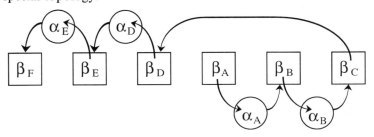

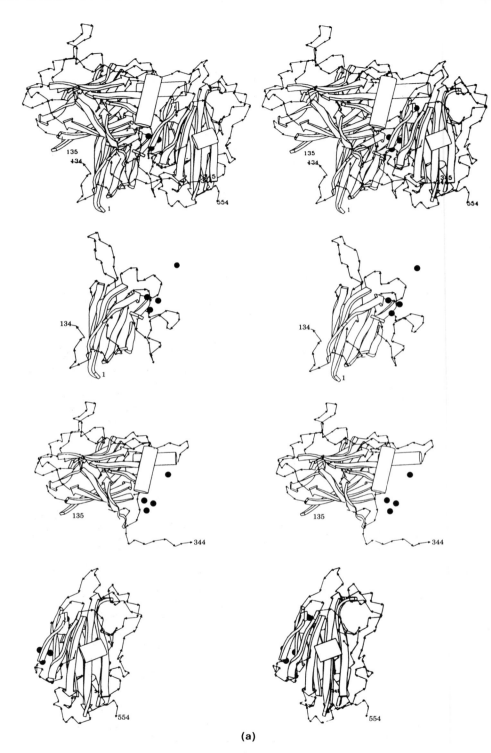

(a)

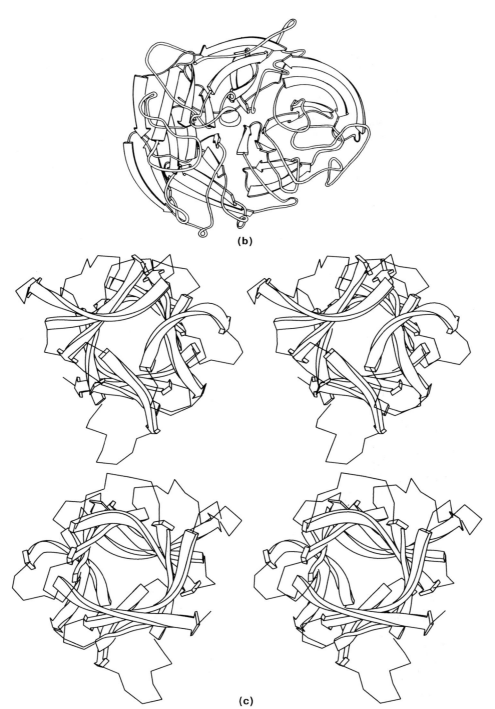

(b)

(c)

Figure 4.16 (a) Ascorbate oxidase. (Courtesy of R. Huber.). (b) Influenza neuraminidase. (Courtesy of P. M. Colman and J. N. Varghese.). (c) Interleukin-1β [2I1B].

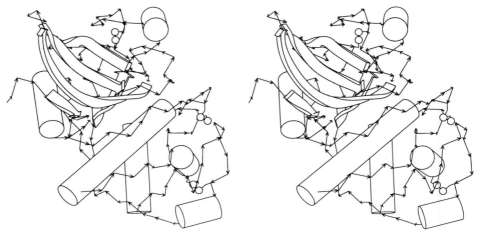

Figure 4.17 Kiwi fruit actinidin [2ACT].

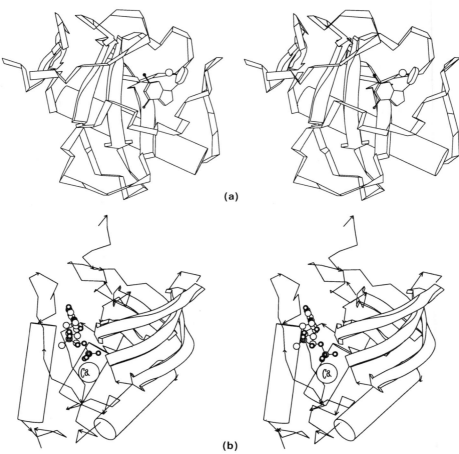

(a)

(b)

Figure 4.18 (a) T1 ribonuclease from *Aspergillus oryzae* [1RNT] with 2'-GMP. (b) Staphylococcal nuclease [2SNS].

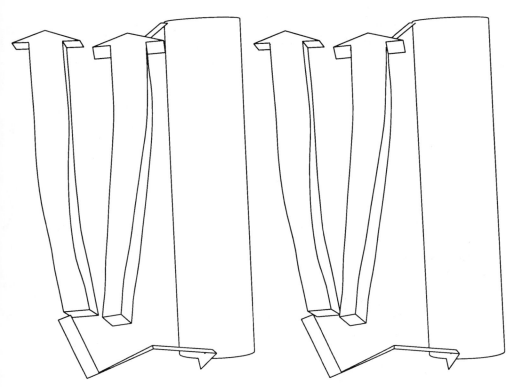

Figure 4.19 The β–α–β unit of tertiary structure [from 6LDH]. (a) schematic representation. (b, c) ball-and-stick models, showing hydrogen-bonding, in different orientations. The orientation in (c) is perpendicular to the sheet. (d, e) The nucleotide-binding domain of dogfish lactate dehydrogenase [6LDH]. In (d) the domain is viewed perpendicular to the sheet. In (e) it is viewed looking down on to the sheet, to show how the long connection between the third and fourth strand naturally creates a cavity adaptable as a binding site (see ref. 66).

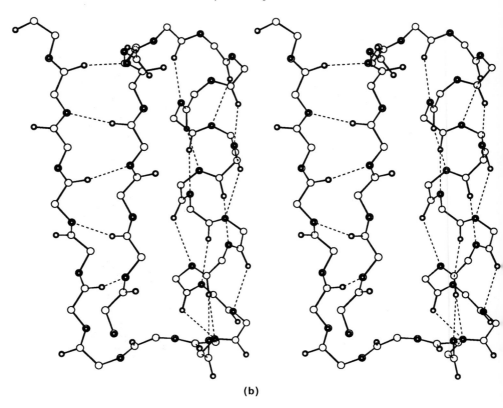

(b)

The long loop between β_C and β_D tends to create a natural pocket for the nucleotide ligand (66). The N–H groups in the last turn of the α_A helix are well-positioned to form hydrogen bonds to phosphate oxygens of the ligand (*Figure 4.20*).

The NAD-binding domain of horse liver alcohol dehydrogenase is typical; other dehydrogenases have very similar domains (see *Figure 4.20*). Flavodoxin and adenylate kinase contain a variation on the theme: they have five strands instead of six (see *Figure 4.21*). Dihydrofolate reductase has eight strands (see *Figure 4.22*).

5.4.3 Closed β–α–β barrel structures

Chicken triose phosphate isomerase is typical of a large number of structures that contain eight β–α–units in which the strands form a sheet wrapped around into a

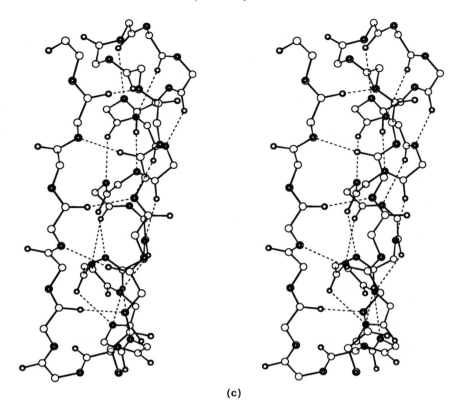

(c)

closed structure, cylindrical in topology. The helices are on the outside of the sheet. (See *Figure 4.23*.) Structures of this type are discussed in Section 6.3.

5.5 Irregular structures

A classification of proteins based on secondary structure must eventually face the structures that contain very few of their residues in helices and sheets. These tend to be stabilized by additional primary chemical bonds. For instance: (1) in the case of wheat germ agglutinin, there are numerous disulphide bridges (*Figure 4.24a,b*). (2) In the case of ferridoxin there are iron–sulphur clusters (*Figure 4.24c*). (3) The 'kringle' structure, first found as a domain of prothrombin but occurring in many other proteins, contains disulphide bridges as well as several short stretches of two-stranded β-sheet (67). (*Figure 4.25*.)

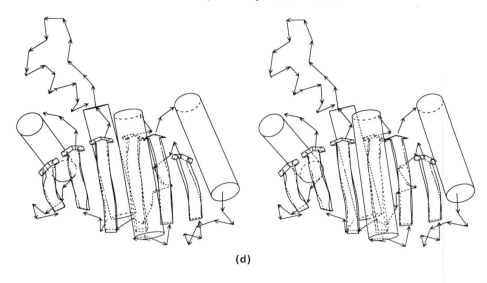

(d)

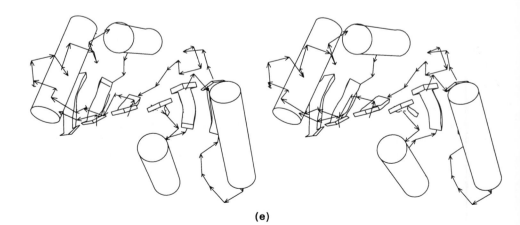

(e)

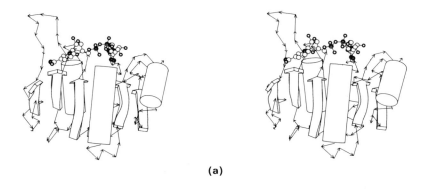

(a)

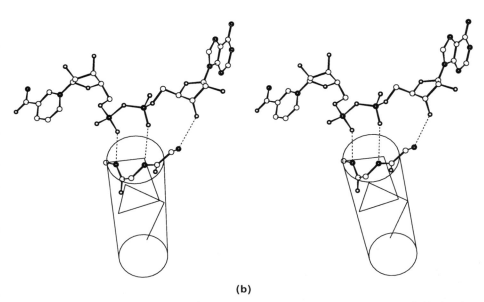

(b)

Figure 4.20 The NAD binding domain of horse liver ADH[6ADH]. (a) The typical fold of such domains, binding NAD. (b) The region around the aA helix, showing hydrogen bonding between the N–H groups in the last turn of the helix and the phosphate oxygens.

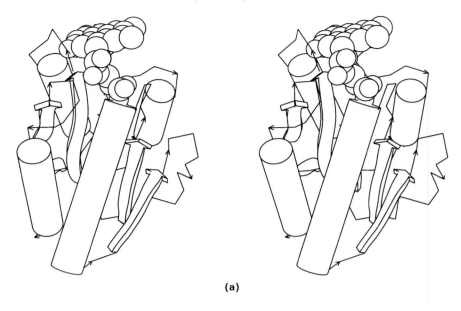

(a)

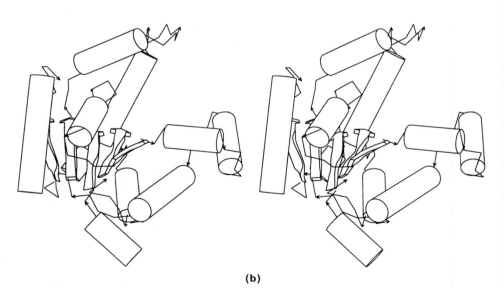

(b)

Figure 4.21 (a) Flavodoxin [3FXN]; (b) Adenylate kinase [3ADK].

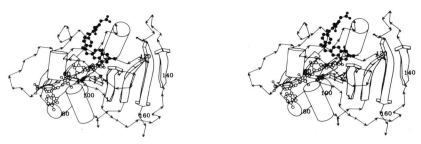

Figure 4.22 Dihydrofolate reductase from *Lactobacillus casei*; tertiary complex with NADPH (open circles) and methotrexate (solid circles) [3DFR].

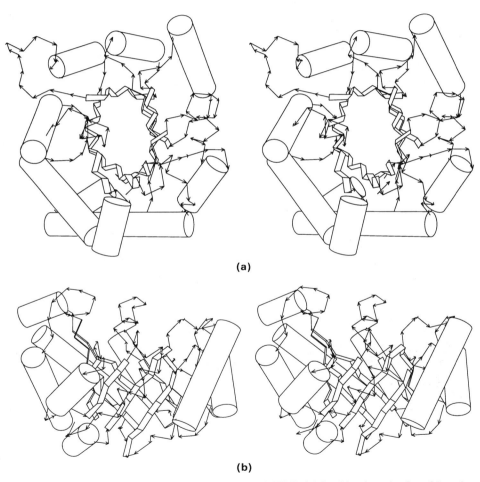

(a)

(b)

Figure 4.23 Chicken triose phosphate isomerase [1TIM]. (a) Looking into the β–α–β barrel; (b) view perpendicular to the barrel.

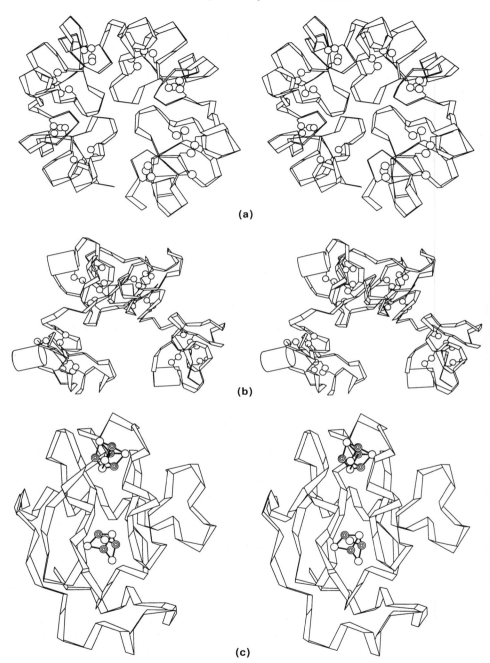

(a)

(b)

(c)

Figure 4.24 (a, b) Wheat germ agglutinin [3WGA]. (c) Ferredoxin from *Azotobacter vinelandii* [4FD1]. This molecule contains one 4Fe–4S cluster and one 3Fe–4S cluster.

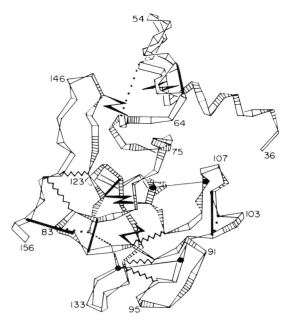

Figure 4.25 The 'kringle' domain from bovine prothrombin. (Courtesy of A. Tulinsky).

6 Tertiary interactions

6.1 Introduction

We have seen, in this chapter, two different sources of similarity among protein structures. Some of these are familial relationships, such as those of the globins. Others reflect the limited ways that secondary structural elements can come together to satisfy the thermodynamic requirements of a stable globular protein structure; including—but by no means limited to—the particular subject of this section: burying of hydrophobic residues in a well-packed interior. These restrictions on secondary structure interactions were discovered by C. Chothia and his co-workers. Here we shall describe helix–helix packings as an example and refer the reader to the original papers on helix–sheet and sheet–sheet interactions.

To examine the packing of residues we must make use of *some* space-filling representation. On an interactive computer graphics device with a colour screen and hardware support for depth cueing and real-time rotation, 'dot-surfaces' are useful; either simply Van der Waals surfaces or the water-accessible surface portrayed by M. Connally's program. (See *Plates 2–3*.) Surfaces are not so effective in a still, black-and-white picture, and it may be more useful to draw serial sections through the Van der Waals envelope of the atoms. These can be

Figure 4.26 The B–G-helix interface in sperm whale myoglobin [1MBO]. (a) Cα-trace: B-helix broken lines, G-helix solid lines. (b) Wire model showing backbone and side-chains in contact in the interface. (c) Serial sections through interface. The packing involves $i \pm 4$ ridges from both helices.

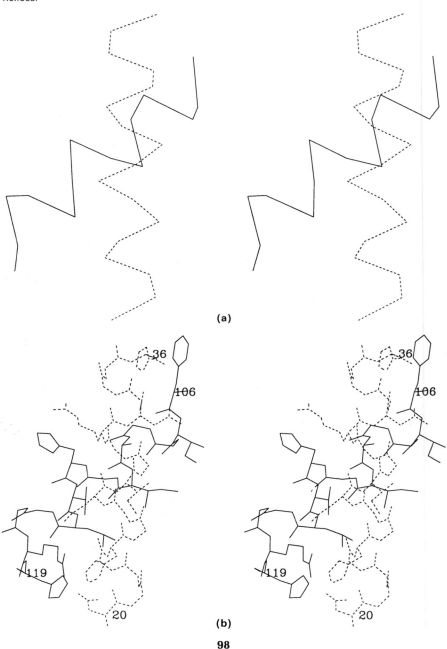

(a)

(b)

combined with skeletal models to assist in the structural interpretation of what is shown. We have seen an example of this in *Figure 3.8*.

6.2 Helix–helix packings

In many proteins, pairs of α-helices are in contact, burying the side-chains in the interface between them. The relative geometry of the helices can be described by the distance of closest approach between their axes, plus the inter-axial angle. Provided that the sides rather than the ends of the helices are in contact, the following geometric properties will hold:

- The interface between the helices will contain a surface patch from each of the helices.

- The shortest line between points on each axis will be perpendicular to both axes and will intersect each axis within the interface patch.

Typically, inter-axial distances are 7–12 Å, with about a 2-Å-inter-penetration of the side-chains. To achieve good packing densities, the two interfaces have complementary surfaces, like the occluding surfaces of (healthy) teeth.

Interaxial angles tend to fall into three classes (41, 42). A closer examination of the nature of the packing at helix–helix interfaces reveals why this is so.

Figure 4.26 shows the interface between the B- and G-helices of sperm whale myoglobin. Residues from the B-helix are shown in solid lines; G-helix residues are shown as broken lines. We can see that the residues from the G-helix form a

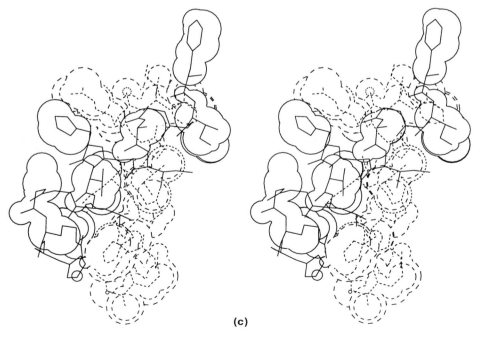

(c)

ridge (created by the side-chains of residues 99, 103, 107) that nestles in a groove between two ridges on the surface of the B-helix (created respectively by residues 19, 23, 27 and 26, 30). In each case the residues are separated by four in the sequence.

If one recalls the basic geometry of the α-helix (3.6 residues per turn) residues separated by 4 in the sequence are close together on the helix surface. Side-chains at these positions are poised fairly naturally to create ridges, which will of course have a defined angle relative to the helix axis. (We will call these $i \pm 4$ ridges.) Thus, in an ideal α-helix, the line joining the Cβ of residue i to the Cβ of residue $i \pm 4$ will make an angle of 26 degrees with the helix axis. (The actual direction of the ridge may deviate moderately from this value because of the shapes and conformations of particular side-chains.) It follows that forming an interface by packing $i \pm 4$ ridges on the two helices will fix the interaxial angle at a value near 50 degrees. This is indeed near the average for one of the classes of interaxial angles.

Although the formation of ridges by residues separated by 4 in the sequence are the most common, ridges can also be formed by residues separated by 3 in the sequence or by residues separated by 1 in the sequence. In the last case the ridge runs nearly around the helix rather than nearly along it. We will call these $i \pm 3$ ridges and $i \pm 1$ ridges, respectively. Which set of ridges is most prominent depends on the relative sizes and the conformations of the various sidechains.

The three classes of helix–helix packings, shown in the three clusters of inter-axial angles in a histogram, correspond to the interaction of different ridges and grooves:

Interacting ridges	Inter-axial angle
$i \pm 4 - i \pm 4$	$-52°$
$i \pm 3 - i \pm 4$	$+23°$
$i \pm 1 - i \pm 4$	$-105°$

Despite the success of the 'ridges-into-grooves' model in rationalizing the main features of the inter-axial angle distribution in packed helices, there are numerous exceptions: pairs of packed helices that have unusual inter-axial angles, and upon close examination are found to have surfaces that do not correspond to the simple 'ridges-into-grooves' packing model. For example, in the B–E-helix packing in sperm whale myoglobin (and other globins) ridges from the helices cross each other, at a notch formed at a pair of glycine residues. (See *Figure 4.27.*) On the one hand, such exceptions 'prove the rule', but on the other they present an intrinsic complication in attempts to use the observed regularities in secondary structure packings to predict structures.

6.3 Packing inside the sheet of *β–α–β* barrel structures

The enzyme triose phosphate isomerase (TIM) has a special type of β–α–β structure. It contains 8 parallel α/β units, the strands forming a β-sheet closed into a cylinder or barrel, and the helices packed against the outside of the sheet (see *Figure 4.23*). Chicken TIM, first solved in 1975, was for a long time the only example, but recently many enzymes containing TIM-like barrels have been found. Over 20 are now known. Like the open β–α–β structures, the active site is at the end of the barrel that corresponds to the C-termini of the strands of sheet. Although they show similar folding patterns, these enzymes catalyse a variety of different reactions, and those with different functions have very dissimilar amino acid sequences: it is not possible to align them by standard techniques.

With many structures available, it became possible to adduce general features of this folding pattern (68):

1. Looking from the *outside*, the sheet is formed by eight parallel strands, tipped by approximately 36 degrees to the barrel axis. (*Figure 4.23b.*) The helices are approximately parallel to the strands, typical of α/β structures (36). The chain always proceeds around the barrel in a counterclockwise direction, viewed from the C-termini of the strands of sheet: locally the chain proceeds 'up' a strand, 'down' a helix, up the next strand, etc. One or two exceptional cases are known, in which a helix is missing [muconate lactonizing enzyme (69)] or a strand is inverted [enolase (70)].

2. A perspicuous view of the *inside* of the structure of the sheet is afforded by 'rolling out the barrel' (*Figure 4.28*). The leftmost (the N-terminal) strand is repeated at the right. To recover the three-dimensional eight-stranded barrel, this diagram must be folded over, and the two images of first strand glued over each other: superposing A on to A' and B on to B' in *Figure 4.28*.

The tipping of the strands to the barrel axis (vertical in *Figure 4.28*) produces a layered structure. Note that the side-chains of each strand point alternately into and out of the barrel; this is an important feature of the packing inside the barrel. The strands vary in length; however, all strands contain three residues at the same height that form a continuous hydrogen-bonded net girding the barrel.

3. The different barrels are similar in topology. McLachlan (71) classified ideal β–α–β barrel topologies. He showed that two *integral* quantities: the number of strands, and the *shear* [= the difference, along the sequence, of the residues forced to correspond when closing the barrel by superposing the two copies of the first strand (see *Figure 4.28*)] determine the tilt of the strands to the barrel axis, the twist of the strands (that is, the average angle between adjacent strands), and the radius of the barrel.

All known β–α–β barrels in proteins have 8 strands, a shear of 8, a tilt of the strands to the barrel axis of approximately 36 degrees, and radii of 6.5–7.5 Å, depending on the eccentricity of the cross-section.

4. The packing of the residues inside the sheet shows a common structural pattern. This is discussed in detail in the following paragraphs.

Glycolate oxidase

Glycolate oxidase (GAO) (*Figure 4.29*) is the most nearly circular in cross-

Figure 4.27 The B–E-helix interface in sperm whale myoglobin [1 MBO]. (a) Cα-trace: B-helix broken lines, E-helix solid lines. (b) Wire model showing backbone and side-chains in contact in the interface. Residue 25 is a glycine. (c) Serial sections through the interface. The ridges cross each other over Gly 25. It is clearly seen that the ridge shown in broken contours is interrupted over Gly 25.

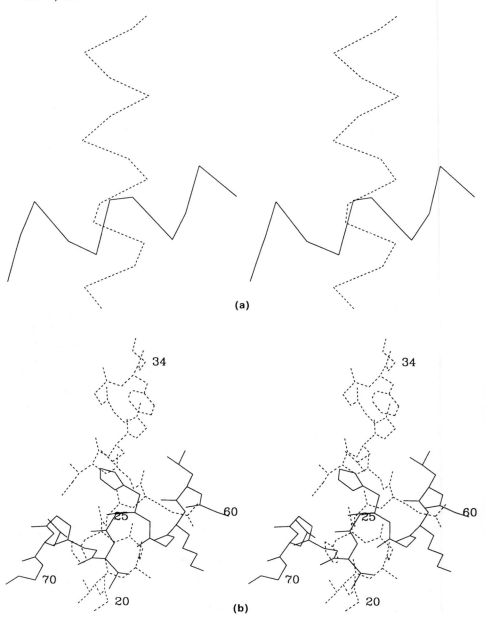

(a)

(b)

102

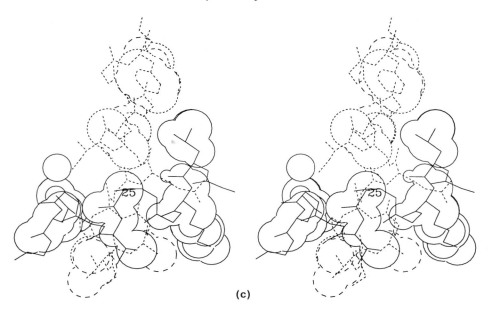

(c)

section, and shows the most symmetrical arrangement of side-chains in the region inside the sheet. It exhibits the general pattern of packing most clearly.

Figure 4.30a shows the hydrogen-bonding net obtained by 'rolling out the barrel' of GAO. On each strand of sheet, alternate side-chains point towards the region inside the sheet and out towards the helices. Twelve residues with letters identifying the side-chains point inwards. The packing inside the barrel is formed by the interactions of these twelve residues. The first, third, fifth, and seventh strands each contribute one side-chain, and the second, fourth, sixth, and eighth strands each contribute two side-chains. Note that residues at the same height along the axis of the sheet (vertical in *Figures 4.28* and *4.30a*) are not nearest neighbours on adjacent strands, because of the tilt of the strands with respect to the barrel axis.

The packing of these side-chains in the barrel interior is shown in *Figure 4.30b–e*. Part *b* is a side view of the sheet of GAO, pruned to three residues per strand. The side-chains occupy three tiers or layers with almost perfect segregation.

The packing of these residues is seen in *Figure 4.30c–e*, which show serial sections through the three layers of GAO. Atoms from odd-numbered strands are outlined by broken lines, and atoms from even-numbered strands by solid lines. In the first and third layers, the four packed side-chains are outlined by broken lines; they belong to odd-numbered strands. In the central layer, the four packed side-chains are outlined by solid lines; these belong to even-numbered strands.

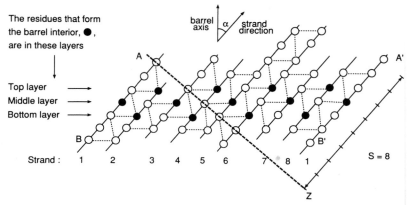

The residues that form
the barrel interior, ●,
are in these layers

↓

Top layer ⟶
Middle layer ⟶
Bottom layer ⟶

Strand :   1   2   3   4   5   6   7   8   1

S = 8

Figure 4.28 Schematic diagram of the hydrogen-bonding nets obtained by unrolling the sheet in chicken triose phosphate isomerase, glycolate oxidase and numerous other proteins containing this type of domain. Each circle represents a residue; broken lines represent hydrogen bonds. Filled circles represent residues in three layers that point into the barrel interior; note their alternation.

This diagram contains nine strands, as the first strand is duplicated at left and right edges. To recover the three-dimensional eight-stranded barrel, the leftmost strand must be superposed on the rightmost by folding the paper into a cylinder and glueing residue A on to residue A' and residue B on to residue B'. This produces the natural barrel, with the strands tipped by 36 degrees to the barrel axis.

To form an eight-stranded barrel with strands *parallel* to the axis, residue A would have to be glued on to point Z. The *shear*, S, was defined by McLachlan (71) as a measure of the stagger of the strands. In this diagram it is the number of residues by which A'—residue on which A is actually superposed—is displaced from Z—the residue on which A would be superposed in a barrel with strands parallel to the axis.

At the centre of the barrel, twelve inward-pointing side-chains pack together. These correspond to the filled circles. Note their arrangement in three parallel layers, in planes lying at the same height relative to the barrel axis. The symmetry of this pattern could accommodate the change of a strand from parallel to anti-parallel, as in enolase (70). This and succeeding figures are from ref. 68.

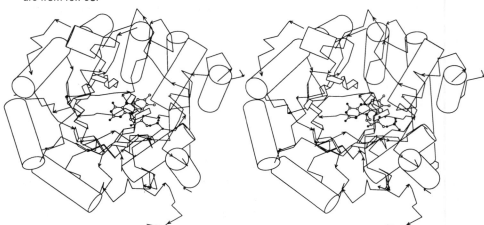

Figure 4.29 Spinach glycolate oxidase, a β–α–β barrel protein [1GOX].

104

These results suggest a simple model for the packing of residues inside the sheet of glycolate oxidase. There is a double alternation similar to a chessboard, but in three dimensions: The tilt of the strands relative to the axis of the sheet, and the twist of the sheet, place the inward-pointing side-chains in layers. Each layer contains four side-chains from alternate strands. The side-chains that point 'in' are on odd-numbered levels on odd-numbered strands, and on even-numbered levels on even-numbered strands. The central region of the barrel is filled by twelve side-chains from three layers. Qualitatively, the packing is a layered *a–b–a* type structure. Successive layers are related by a rotation by 45 degrees around the barrel axis and a translation along the axis by approximately 3 Å.

Triose phosphate isomerase
The barrel of chicken TIM has a very non-circular cross-section; it also shows significant deviation from the paradigm of the packing. Of the twelve residues in chicken TIM the side-chains of which are expected to form the barrel interior, three are glycines and two are alanines (*Figure 4.31a*). These five small residues reduce the volume of sidechains available to pack the interior, and produce substantial distortion; nevertheless, the structure can still be understood in terms of the same basic pattern of residue packing as in GAO.

Like GAO, TIM contains three layers of side-chains packed inside the sheet. *Figure 4.31a* shows the unrolled β-sheet of TIM, and *Figure 4.31b–e* shows the packing inside the sheet.

In the top layer, the cavity next to 42 Gly is filled by the side-chain of 11 Asn, from above (*Figure 4.31c*). The side-chain of this residue is buried, and its position is stabilized by hydrogen bonds. In the central layer, the cavity next to 9 Gly is partly filled by the backbone movement that produces the non-circular cross-section of the barrel (*Figure 4.31d*). As this is only partly successful, the structure has a cavity adjacent to this residue (*Figure 4.31d*); such large cavities are rare but not unknown in protein structures (40). In the bottom layer, the region that 228 Gly would be expected to occupy is filled by the side-chain of 7 Phe, from below (*Figure 4.31e*).

In summary, the packing of the region inside the sheet of GAO and TIM can be understood in terms of a three-tiered arrangement in which alternate strands contribute side-chains to alternate layers. When interior residues are small (glycines or alanines) part of a residue on an adjacent layer or a residue from outside the barrel inserts its side-chain into the layer; or the backbone may be deformed.

Classes of β–α–β barrels—evolutionary considerations
Underlying the similarity of folding of α/β-barrels lurk *two* topologically distinct classes. In the girdle of three residues per strand common to the structures, alternate strands contribute one or two side-chains, respectively, to the region within the sheet. In one class of structures the odd-numbered strands contribute one residue and the even-numbered strands two residues. In the other class, the

Figure 4.30 Structural analysis of the region inside the barrel of spinach glycolate oxidase (GAO).

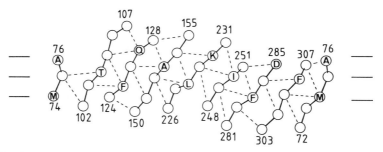

(a) The β-sheet of GAO, drawn by unrolling the barrel. Each circle represents a residue; one-letter code identify residues the side-chains of which pack inside the barrel. Broken lines represent hydrogen bonds. Numerals indicate residue numbers. Nine strands are shown: the edge strand (residues 72–76) is duplicated [1GOX].

This figure and *Figure 4.31a* are cylindrical projections drawn from atomic coordinates and therefore give an accurate picture of the residue positions. In contrast, *Figure 4.28* is idealized.

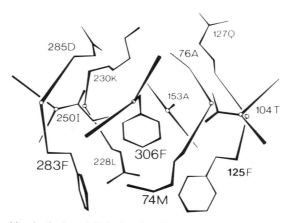

(b) Residue packing in the barrel. This drawing shows the eight strands of the GAO β-sheet, pruned to the three residues per strand, and the inward-pointing side-chains. The view is perpendicular to the barrel axis.

Plates

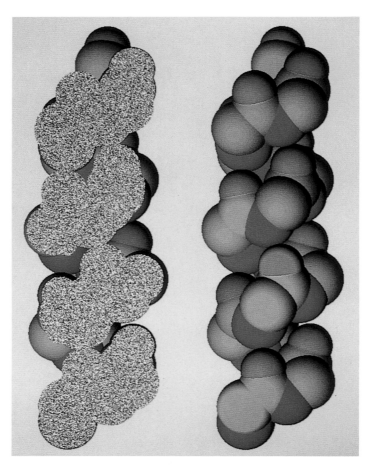

Plate 1. 'Cheese-wire' cut through a space-filling model of an α-helix.

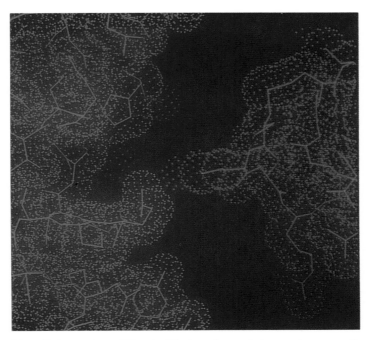

Plate 2. Interaction of Van der Waals surfaces of the trypsin–pancreatic trypsin inhibitor complex [1TPA]. (Courtesy of A. Tramontano.)

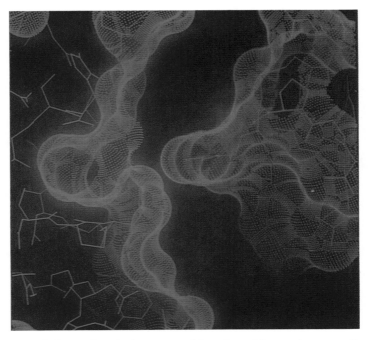

Plate 3. Interaction of solvent-accessible surface of the trypsin–pancreatic trypsin inhibitor complex [1TPA]. (Courtesy of A. Tramontano.)

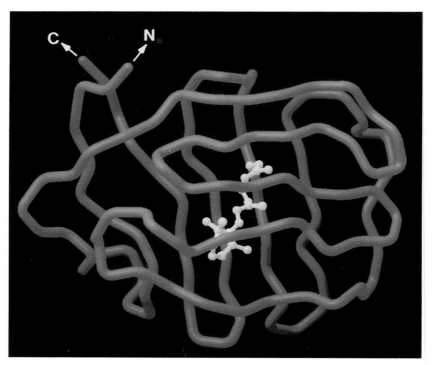

Plate 4. Photoactive yellow protein. (Courtesy of D. McRee, M. Pique, J. Tainer, and E. Getzoff; see ref. 123.)

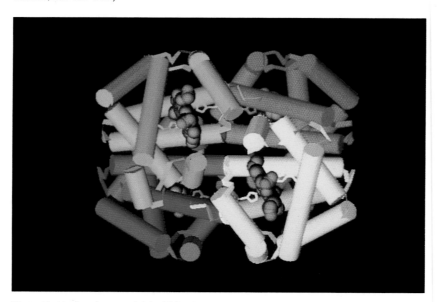

Plate 5. Human haemoglobin. This picture appeared on a poster for the exhibit at the Science Museum, South Kensington, commemorating the fortieth anniversary of the founding of the MRC Laboratory of Molecular Biology.

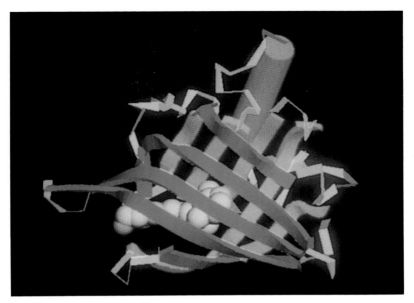

Plate 6. Retinol-binding protein. (Courtesy of T. A. Jones.)

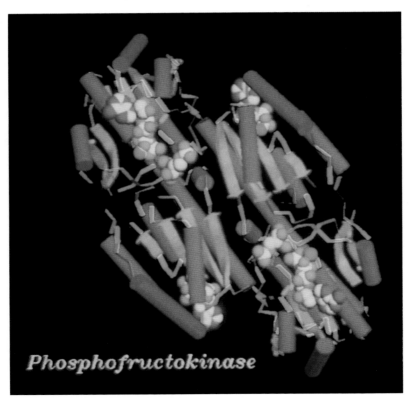

Plate 7. Phosphofructokinase.

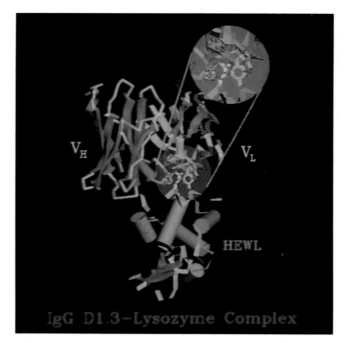

Plate 8. Antigen–antibody complex: D1.3 and hen egg white lysozyme. (Courtesy of S. E. V. Phillips and R. J. Poljak.)

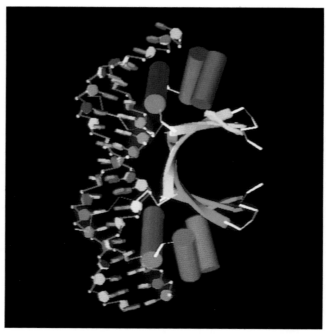

Plate 9. Binding of λ-cro to DNA. (Courtesy of B. W. Matthews.)

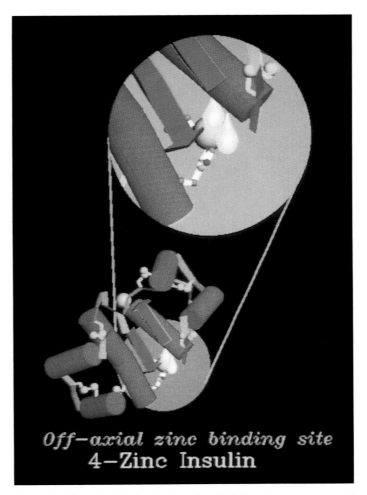

Plate 10. The off-axial zinc binding site of 4Zn Insulin.

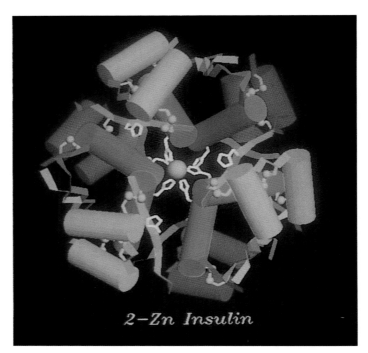

Plate 11. The 2Zn insulin hexamer.

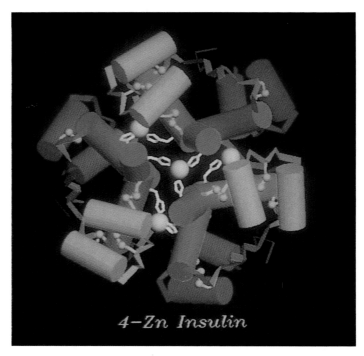

Plate 12. The 4Zn insulin hexamer.

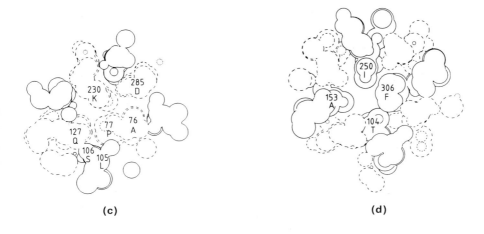

(c)   (d)

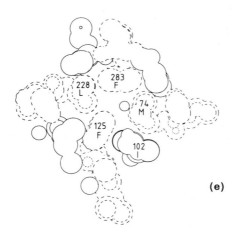

(e)

(c, d, e) Serial sections cut through a space-filling model (Van der Waals slices) of the three layers of residues packing inside the barrel of spinach glycolate oxidase. In each drawing three slices separated by 1 Å are shown. Atoms from alternate strands are drawn by solid and broken lines.

The reader is urged to follow each of the twelve residues from part (a) to part (b) to part (c), (d), or (e) of this figure.

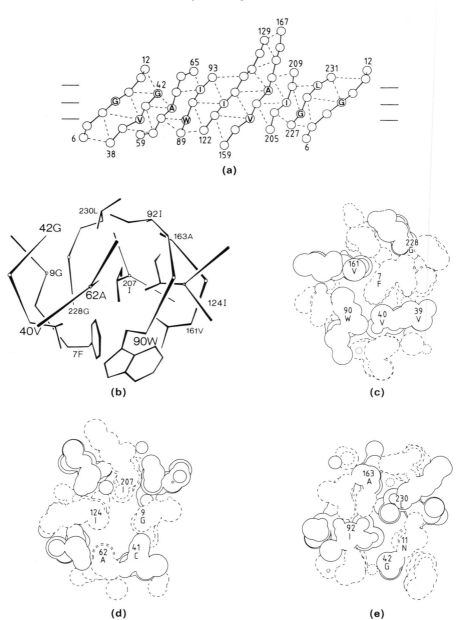

Figure 4.31 Chicken triose phosphate isomerase (TIM). (a) The β-sheet of TIM, drawn by unrolling the barrel, and represented analogously to *Figure 4.30a* [1TIM]. (b) Residue packing in the barrel of TIM. All but the first strand are pruned to 3 residues per strand; the first strand includes residue Phe 7, the side-chain of which fills a cavity next to Gly 228. (c, d, e) Van der Waals slices through the three layers of residues packing inside the sheet of chicken TIM.

odd-numbered strands contribute two residues and the even-numbered strands one residue. GAO and TIM are in different classes (compare *Figure 4.30a* and *4.31a*).

Are the β–α–β proteins in these two classes related by evolution? If structures in these two classes did not arise independently, there must be a pathway between them that evolution could follow (51, 72). The simplest mechanism by which structures in these two classes could be interconverted would involve an intermediate structure with two or four layers packing in the barrel interior.

6.4 Folded protein structures as solved jigsaw puzzles

The demonstration that tertiary structural interactions tend to be subject to geometric constraints impresses a form on the original 'oil drop' model of protein interiors. A native protein structure depends on a mutually-consistent set of good packings among the elements of secondary structure that interact.

The model that emerges for a native protein is that of a three-dimensional jigsaw puzzle. However, there is this difference: the pieces of the puzzle only acquire rigidity by virtue of their interactions. The events that take place during the folding process remain obscure. As in the crystallization of a short peptide, the elements must somehow explore different individual shapes, and different relative positions and orientations, to find the proper shapes and the proper fit.

Thus, the native state of a globular protein is similar to an assembled jigsaw puzzle. But in the unfolded state the pieces are not only taken apart, their shapes are changed and disguised.

7 Loops

The term *loops* refers to sections of the polypeptide chain that connect regions of secondary structure. Frequently, helices and strands of sheets run across a protein or domain from one surface to another, and loops are characterized by (a) appearing on the surfaces of proteins, and (b) reversing the direction of the chain. A typical globular protein contains two-thirds of its residues in helices and sheets, and one-third in loops. [These statements are common observations, but not hard-and-fast rules (see refs 73 and 74).]

Some loops connect successive helices or strands of sheet that interact with each other to form paradigmatic supersecondary structures, such as a pair of anti-parallel packed helices, a β-hairpin (that is, two successive strands of anti-parallel sheet connected by a turn) or a strand of sheet followed by an *anti-parallel* helix followed by a parallel strand of β-sheet, as seen in the typical nucleotide-binding domain and in β–α–β barrels. The last case requires two loops.

In many enzymes, loops contain functional residues. Because loops tend to be more flexible in conformational changes than helices and sheets, they are often used when a protein needs to respond to changes in state of ligation, as in lactate dehydrogenase or triose phosphate isomerase (*Figure 4.32*). [Allosteric changes

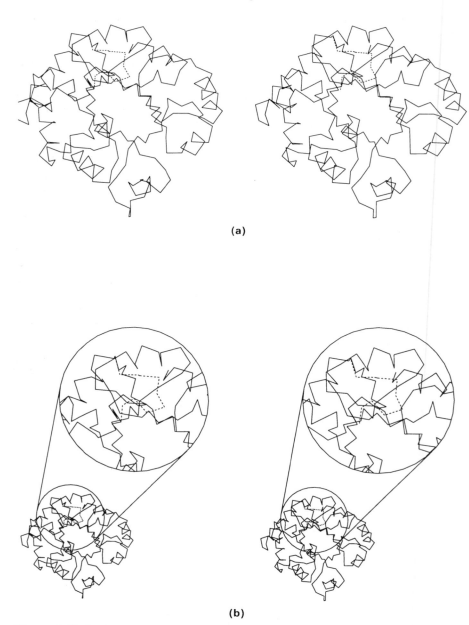

(a)

(b)

Figure 4.32 Conformational change in trypanosome triose phosphate isomerase upon substrate binding. (Courtesy of R. Wierenga and M. Noble.) Solid line: unligated state; broken line: region of major conformational difference in ligated state. (a) The general structure is a β–α–β barrel; the binding site is at the mouth of the barrel at the C-termini of the strands. (b) A 'blow-up' of the region of conformational change.

involve larger-scale changes in relative geometry and packing of entire subunits (See Chapter 5).]

Other loops that have a purely structural role, and that appear on the surfaces of proteins, change fairly rapidly in evolution. They are frequently subject to insertions and deletions, as well as to amino acid substitutions. From a set of aligned sequences of related proteins it is possible by analysis of hydrophobicity profiles and patterns of sequence change to infer the positions of loops in a protein of unknown structure fairly well (75–77).

Hairpin loops (those that connect successive strands of anti-parallel β-sheet) have been studied extensively to classify them and to elucidate the determinants of their conformations (refs 72, and 78–83).

For a short region of polypeptide chain only 3–4 residues in length to reverse direction, and fold back on itself to form a hairpin, a residue in a conformation in the non-allowed region of the Sasisekharan–Ramachandran diagram is generally required. Therefore the conformations of short loops depend primarily on the position within the loop of special residues (usually Gly, Asn, or Pro) that allow the chain to take up an unusual conformation. Gly and Asn are often in the α_L conformation. The peptide bonds preceding Pro are often *cis*. As first pointed out by Sibanda and Thornton (79), the conformation of a short hairpin can often be deduced from the position in the sequence of this special residue. Thus, *Figure 4.33* shows the hairpin of sequence Glu-Gly-Gly-Val from actinidin; compare *Figure 4.34* showing the sequence Ser-Gly-Ser-Ser from elastase.

Not all loops in proteins are short, however. Leszczynski and Rose (84) pointed out that proteins often contain loops 6–10 residues in length; and some

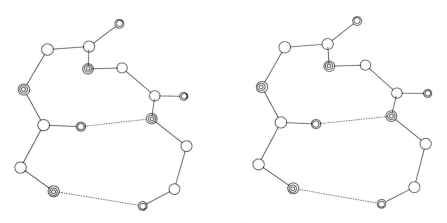

Figure 4.33 A Type I' hairpin from actinidin [2ACT], with the sequence:

Gly—Gly
| |
Glu=Val

(where the double lines indicate the two β-sheet hydrogen bonds between the Glu and Val residues).

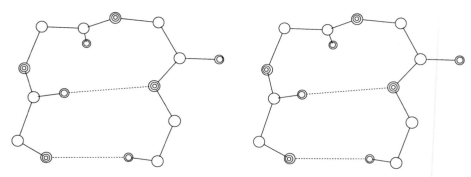

Figure 4.34 A Type II' hairpin from Elastase [3EST], with the sequence:

$$\begin{array}{c} \text{Gly—Ser} \\ | \quad | \\ \text{Ser=Ser} \end{array}$$

loops are much longer. The determinants of the conformations of such loops are not intrinsic to the amino acid sequence of the loop itself, but involve tertiary interactions: hydrogen-bonding and packing. One way to try to understand them is to compare the structural contexts of loops of similar conformation in unrelated proteins. In these cases, the structural similarity is not the result of evolutionary relationships. Jones and Thirup (85) have developed methods to identify regions of similar conformation by searching the database of protein structures.

Two interesting principles have emerged from such studies, and will be illustrated in the examples that follow (86):

1. Medium-sized loops of similar conformation in unrelated proteins are often stabilized by similar interactions.

2. The loops may appear, however, in entirely different structural contexts, and the surroundings of the loops, that provide these interactions, may be constructed in different ways in different proteins.

Example 1. For medium-sized loops that form compact structures, the major conformational determinants are hydrogen bonds to the inward-pointing main-chain polar atoms of the loop.

Figure 4.35 illustrates a loop from a immunoglobulin domain. Its conformation is determined mainly by the *cis*-peptide preceding the Pro residue at the right, and hydrogen bonds formed by the side-chain of the Asn residue at the N-terminus of the loop.

A loop of very similar conformation occurs in tomato bushy stunt virus, 2TBV (*Figure 4.36*).

These loops have different structural contexts: In immunoglobulins, it is a hairpin; in the virus it connects strands from different sheets (*Figure 4.37*). Nevertheless, the structural determinants of the virus loop are very similar to

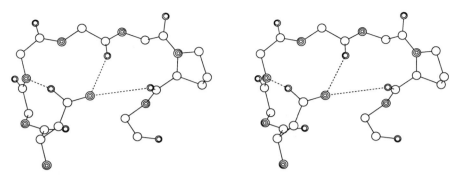

Figure 4.35 A loop from the antigen-binding site of a typical immunoglobulin V_κ domain. This is the third hypervariable loop of the light chain [1MCP]. This and succeeding figures from ref. 86.

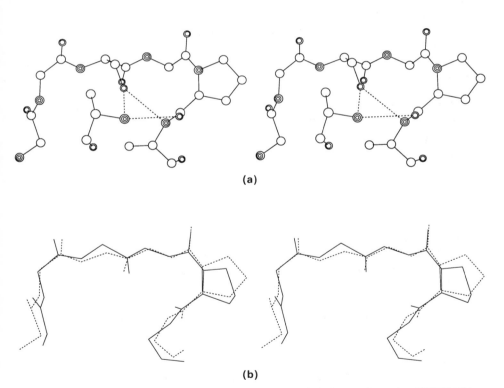

(a)

(b)

Figure 4.36 (a) A loop from tomato bushy stunt virus, of similar conformation [2TBV]. (b) Superposition of loop from immunoglobulin (solid lines) and virus (broken lines).

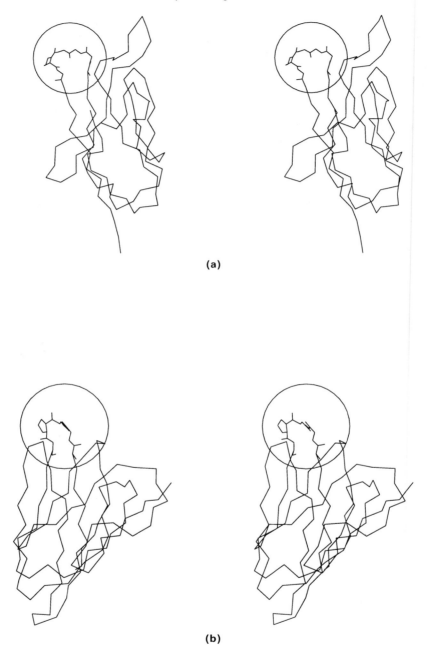

(a)

(b)

Figure 4.37 Difference in structural context between (a) loop in immunoglobulin, a hairpin linking successive antiparallel strands in one β-sheet [1MCP], and (b) loop in virus, linking strands of two β-sheets [2TBV].

those of the immunoglobulin loop. There are *cis*-prolines at equivalent positions. Hydrogen bonds to the main chain of the loop, similar to those made by the Asn side-chain in the immunoglobulin domain, are made in the virus by the carbonyl oxygen of an alanine, distant in the sequence, but occupying the same position in space relative to the loop.

A related conformation occurs in a hairpin in cytochrome c_3 from *Desulfovibrio vulgaris*, 2CDV (*Figure 4.38*). There is no *cis*-proline: the corresponding residue is Gln, and the peptide bond is *trans*. This, however, produces only a local distortion of the loop. Again there are similarities in the stabilizing interactions. In cytochrome c_3, hydrogen bonds are formed by one of the propionyl groups of a haem. Its carboxyl group occupies the same region of space relative to the loop as the amide group of the Asn of the immunoglobulin domain, or the main chain carbonyl of the virus.

It appears that the conformation of the loop makes certain demands on the rest of the protein for specific interactions. Different proteins can meet these demands in a variety of ways.

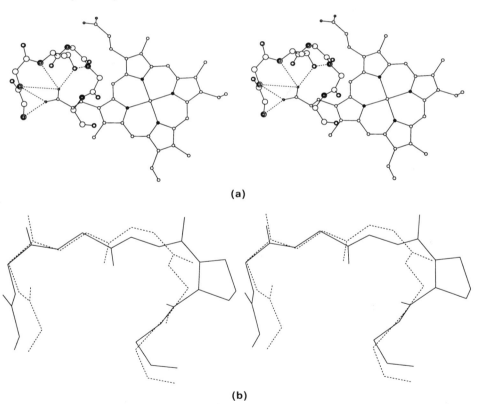

(a)

(b)

Figure 4.38 (a) A loop from cytochrome c_3, with a related conformation [2CDV]; (b) superposition of loop from immunoglobulin (solid lines) and cytochrome c_3 (broken lines).

Example 2. *For loops not compact in themselves, which link more distant secondary structures, the major determinant of the observed conformation is the packing of residues against the rest of the protein.*

Figure 4.39 shows another loop from the antigen-binding site of immuno-globins. This loop connects strands in two different β-sheets. Typically, the chain is coiled to form a distorted helix, and a hydrophobic side-chain is buried in a cavity between the sheets.

A region from the insect globin, *Chironomus* erythrocruorin, has the same conformation as this immunoglobulin loop (*Figure 4.40*). In the globin, the loop links two helices which pack together with their axes approximately anti-parallel. As in the immunoglobulin loop, the residue in the fourth position is a Phe buried in a cavity. This residue appears to be a kind of 'nodal point' of the loop: it requires correct positioning and is the focus of interactions.

How does nature form cavities to pack the same or similar side-chains from very different structural elements: sheets in the immunoglobulins and helices in erythrocruorin? In immunoglobulins the Phe side-chain is packed between four sets of residues: these include the two strands linked by the loop plus additional material from elsewhere in the chain (*Figure 4.41a*). In erythrocruorin, the Phe also has four sets of neighbours, arising from ridges formed by side-chains adjacent on the helix surfaces (*Figure 4.41b*). In both cases the elements of secondary structure linked by the loop contribute to the packing of the buried residue; in immunoglobins these are supplemented by other residues.

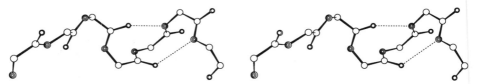

Figure 4.39 A loop from the antigen-binding site of a typical immunoglobulin V$_H$ domain. This is the first hypervariable loop of the heavy chain of Fab KOL [1FB4].

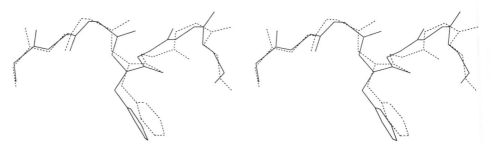

Figure 4.40 Superposition of loops similar conformation from immunoglobulin (solid lines) [1FB4] and *Chironomus* erythrocruorin, a globin (broken lines) [1ECO]. Note the similarity in position of the phe side-chains.

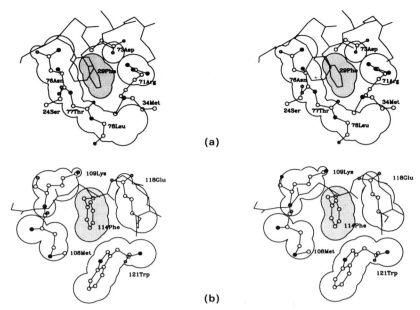

(a)

(b)

Figure 4.41 Packing of phe side-chain around loops of similar conformation in (a) immunoglobulin, (b) erythrocruorin. In each case, the Van der Waals envelope of the phe is shaded.

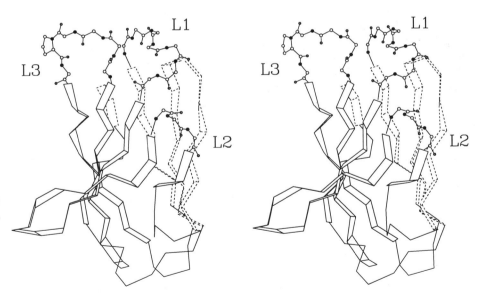

Figure 4.42 A typical immunoglobulin domain (V_L REI) showing the three hypervariable loops [1REI].

As in the previous example, the interactions appear similar from the point of view of the loop, but the proteins create the surroundings out of different material. Or, in other words, the conformation of the loop dictates the stabilizing interactions, but different proteins may use different topological arrangements in the surroundings of the loops to achieve the correct relative positioning of the residues that provide them.

7.1 Antigen-binding sites of immunoglobulins

Immunoglobulins are built of domains with a common fold: two β-sheets packed face to face (*Figure 4.42*). The antigen-binding site contains six loops, three from the variable domain of the light chain (V_L) and three from the variable domain of the heavy chain (V_H). (See *Plate 8.*) These loops are denoted L1, L2, and L3 and H1, H2, and H3, respectively. L2, L3, H2, and H3 are hairpins. L1 and H1

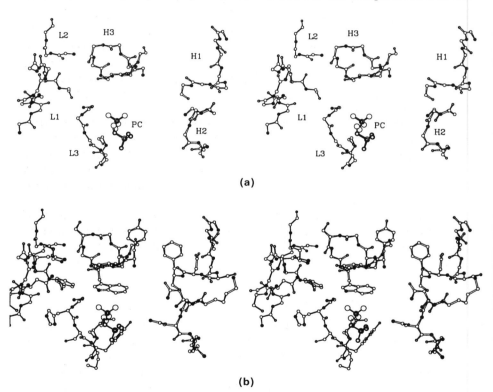

(a)

(b)

Figure 4.43 Binding site of Fab McPC603, showing six antigen-binding loops [2MCP] and the hapten phosphorylcholine (PC). The light chain is drawn with solid bonds; the heavy chain with hollow bonds. The three loops from the light chain are L1, L2, and L3. The three loops from the heavy chain are H1, H2, and H3. Note the very rough dyad symmetry in the placement of the loops: H1 opposite L1, H2 opposite L2, H3 opposite L3. (a) backbone only; (b) backbone plus side-chains.

connect strands of different sheets within V_L and V_H domains. (Can the reader spot the loops similar to those discussed in the previous section?)

Variations in the lengths and amino acid sequences of these loops are primarily responsible for the differences in specificity and affinity of different immuno-globulins. Some of the hairpins are short, and their conformation usually follows the rules relating sequence to structure in short hairpins. (However, cases are known where other interactions override the predisposition of the sequence to a particular conformation of the loop.) Other loops are medium-sized and are described by the principles discussed in the preceding examples. These structures are stabilized by a small fraction of the residues, through packing or hydrogen bonding interactions, or the ability to assume unusual mainchain conformations. Other residues are relatively free to vary, to modulate the surface topography and charge distribution of the antigen-binding site.

For at least five of these six loops, there is a discrete and relatively small repertoire of conformations called 'canonical structures' (72, 81, 87) (*Figure 4.44*). The common appearance in other immunoglobulin sequences of the specific patterns of residues responsible for the conformations of the canonical structures shows that the loop conformations we see in the relatively few known crystal structures occur rather generally. This makes it possible to predict much of the structure of antigen-binding regions of immunoglobulins from the amino acid sequences of their variable domains.

In addition to predictions based on the analysis of conformational determinants of loops in known structures, procedures based on exhaustive exploration of conformation space, and on the selection of a model by conformational energy calculations, are also showing promise (see ref. 88). Perhaps the ultimate success for this special but important problem will come from a combination of all these methods (see ref. 89).

The possibility of accurate prediction of the three-dimensional structures of antigen-binding sites of immunoglobulins is of great importance for engineering antibodies with prescribed specificities, for both scientific and clinical applications.

8 Protein–ligand interactions

Many protein structures contain inorganic ions, water, or other inorganic or organic molecules. These may be intrinsic parts of the structures, as in the case of the zinc ions of insulin or the haem groups of globins and cytochromes c; or they may be substrates, inhibitors, effectors, or antigens in the case of antibodies, which interact with a structure that is fully native even in their absence. Some ligands may themselves be proteins; for example, bovine pancreatic trypsin inhibitor in its complexes with serine proteases, or hen egg white lysozyme in its complexes with immunoglobulins. Others may be nucleic acids, such as the oligonucleotides bound to repressors, tRNA bound to aminoacyl-tRNA synthetases, or the genomes of viruses enfolded within coat protein aggregates.

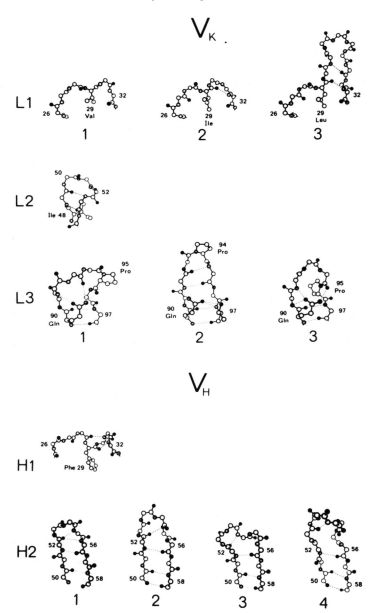

Figure 4.44 Canonical structures of five hypervariable loops of immunoglobulins. The sixth loop, H3, is more variable in length and conformation. (See ref. 86.)

Usually, ligands participate directly in the function of a protein, and are therefore among the most interesting components of a structure to study. The environment within the protein that surrounds and interacts with the ligands is in many cases equivalent to the active site. Determination of the structure of a protein with bound substrate/inhibitor/co-factor is often a necessary condition for elucidating the mechanism of function.

In some cases, ligand binding produces little conformational change in the protein. Sperm whale myoglobin is an example of this: the oxy and deoxy structures can be superposed almost exactly (see Section 3.1 in Chapter 5). In other cases, change in state of ligation produces either simple or profound changes in protein structures. Such a response is shown by insulin, which undergoes a large conformational change between the two-zinc and four-zinc forms, and proteins such as haemoglobin, aspartate transcarbamylase, phosphofructokinase, and phosphorylase which undergo allosteric transitions. Two facts are obvious but must none the less be emphasized: (1) The only way to define the conformational change that accompanies ligand binding (or to show that none occurs) is to solve the structure of the protein in both states of ligation. (2) The fact that a protein structure remains the same in two states of ligation does not mean that the *transition* between the two states does not require mobility in the protein conformation. In the case of sperm whale myoglobin, for example, there is no free channel for the oxygen molecule into its binding site in the haem pocket in the static structures given by X-ray crystallography. (Indeed it is rare that X-ray crystallography can provide information about the pathways of conformational transitions.)

8.1 Illustrating protein–ligand interactions

What kinds of pictures should we draw to help elucidate protein–ligand interactions?

Sometimes, the ligand and the whole active site itself are relatively small in their entirety, and can be intelligibly depicted by fairly simple representations. A complication may arise even in this case if it is desired to show both the details of the active site and its relationship to the overall structure. For this, it is sometimes useful to show the entire molecule schematically, together with a 'blow-up' of the active site. *Figure 4.45* illustrates this technique, using as examples the copper-binding site of poplar leaf plastocyanin, and the binding of sulphate to a transport protein of *S. typhimurium*.

In the cases of proteins that undergo ligand-induced conformational changes, one has the simultaneous problem of showing the site and the conformational change. Although this problem can in principle be solved by a combination of the techniques described in this and the preceding chapter, it is often wiser not to try to show everything in one picture.

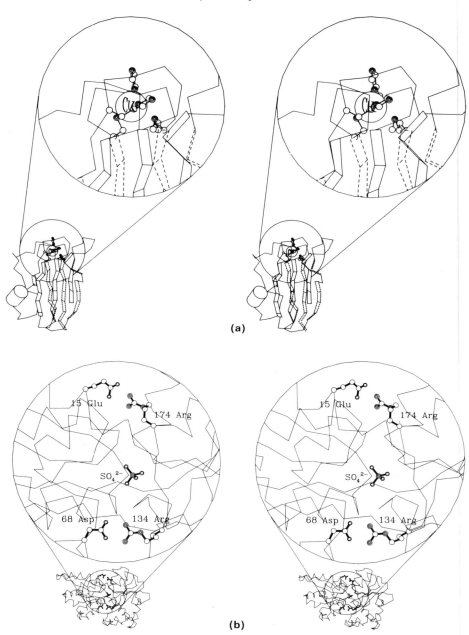

Figure 4.45 (a) 'blow-up' of the copper-binding site of poplar leaf plastocyanin, showing both the details of the ligation and the relationship of the binding site to the overall structure of the molecule [1PCY]. (b) A 'blow-up' of the sulphate-binding site in a transport protein of *S. typhimurium*. (Courtesy of F. A. Quiocho; see ref. 90.)

8.2 Organic ligands

Many protein structures contain co-factors, inhibitors, substrates, or effectors; we have already seen some examples. *Figure 4.46* shows hen egg white lysozyme binding a trisaccharide inhibitor.

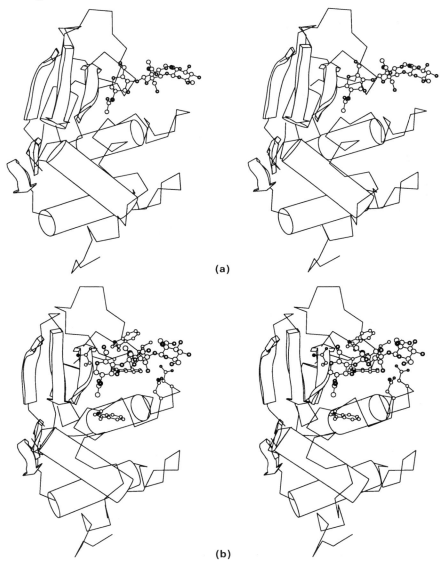

(a)

(b)

Figure 4.46 Hen egg white lysozyme, binding oligosaccharide. (Courtesy of J. Cheetham and Sir D. C. Phillips.) (a) Ball-and-stick representation of inhibitor only; (b) the side-chains in contact with the inhibitor are shown also, with smaller atomic radii.

Plate 6 shows retinol-binding protein enclosing the retinol molecule.

Figure 4.47 and *Plate 7* show a dimer of the allosteric enzyme phosphofructoki-nase from *B. stearothermophilus*, containing the substrate, fructose-6-phosphate, the co-factor ATP and the effector, ADP. (See also Section 5.4.2.)

Plate 8 shows the complex between the monoclonal antibody D1.3 and its antigen hen egg white lysozyme.

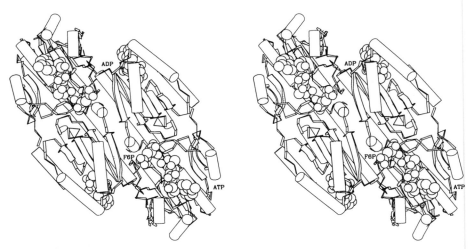

Figure 4.47 Phosphofructokinase, R state, from *B. stearothermophilus*, showing substrate fructose-6-phosphate F6P, co-factor ATP, and effector ADP. The molecule is a tetramer, of which only two subunits are shown [1PFK].

8.3 Nucleic acids

A schematic representation of nucleic acids, developed by V. I. Lesk and the author, is useful in illustrating polynucleotides themselves, and the binding of polynucleotides to proteins (91).

Figure 4.48 and *Plate 9* show the model of the binding of DNA to λ-cro. *Figure 4.49* shows the interaction of the RNA and coat protein in tobacco mosaic virus.

8.4 Water molecules

In the X-ray crystal structure analysis of proteins, it is not possible to see hydrogen atoms in electron-density maps. (This is because the scattering power of an atom for X-rays depends on the number of electrons it contains.) Water molecules that are well-ordered present themselves as single peaks at the position of the oxygen atoms. It requires a high-resolution structure determination, and nice judgement, to assign the positions of water molecules in an electron-density map. Many crystallographers adopt the criterion that a peak in the electron

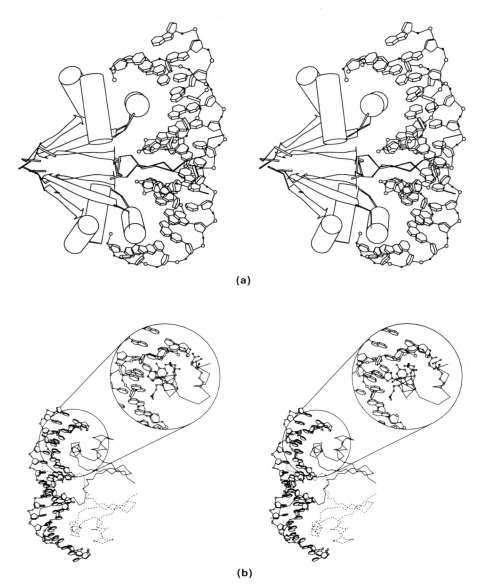

Figure 4.48 (a) Model of the binding of DNA and λ-cro; (b) 'blow-up' showing helix-turn-helix DNA binding motif [1CRO]. (Courtesy of B. W. Matthews.)

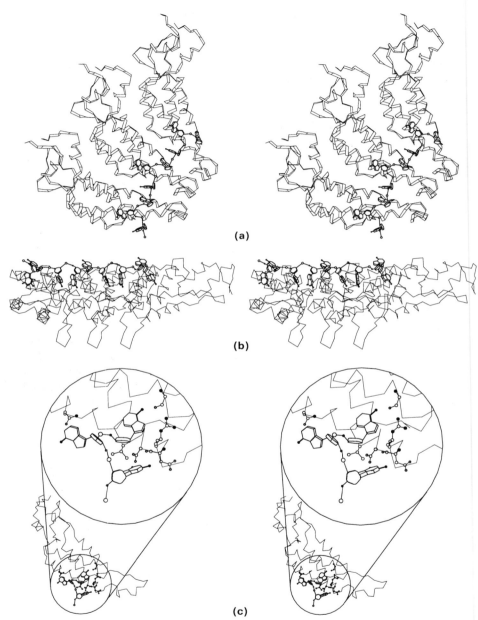

Figure 4.49 Interaction of RNA and coat protein in tobacco mosaic virus [2TMV]. In parts (a, b) three successive subunits and the associate nine ribonucleotide units are shown. (a) view parallel to axis of the virus. (b) View perpendicular to axis of virus. (c) 'Blow-up' of one subunit showing the three associated ribonucleotide units and protein sidechains in contact with the nucleic acid.

density distribution will be accepted as a water molecule only if it is in a position at which a water molecule could make at least one reasonable hydrogen bond.

Unambiguous identification of water molecules is possible using neutron diffraction. The relative scattering of neutrons by hydrogen is greater; and it is the nuclei that are responsible for the scattering, so withdrawal of electrons from the hydrogen atoms to the more electronegative oxygen atom does not alter the effective shape of the molecule. As a result, water molecules can be seen as bent triatomic molecules, and their positions and orientations identified unambiguously. (Currently, the Protein Data Bank contains neutron structures of myoglobin, trypsin, bovine pancreatic trypsin inhibitor, and ribonuclease A, and a bibliographic entry for a neutron study of lysozyme.)

In many protein structures, water molecules clothe much of the surface of the protein. Other waters play specific roles in the structure. For example, the off-axial zinc-binding site in four-zinc insulin is formed by two histidine residues and two water molecules (*Plate 10*). The 'iconic' representation of the water molecules as teardrop shapes distinguishes them conveniently from other atoms. Buried water molecules are rare, but are integral parts of the structures of some proteins.

Water molecules are sometimes found at crucial positions at interfaces between domains or subunits. *Figure 4.50* shows the participation of three water molecules between the coat protein monomers in the disc of tobacco mosaic virus. (The disc is an aggregate of 34 monomers believed to be an intermediate in virus assembly.) Figure *4.51* shows a β-sheet in *Escherichia coli* phosphofructokinase in two allosteric states. In the T state, the β-sheet is continuous across a monomer–monomer interface. In the R state the monomers have moved apart and water molecules are inserted between them, bridging the separated strands of β-sheets (92).

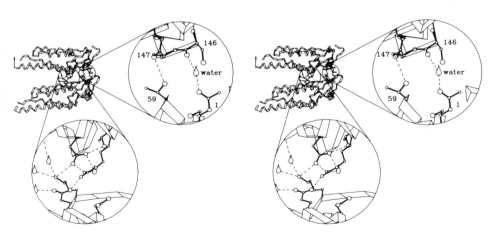

Figure 4.50 Interactions at an interface between subunits of the tobacco mosaic virus disc. (Picture by and courtesy of A. Mondragon.)

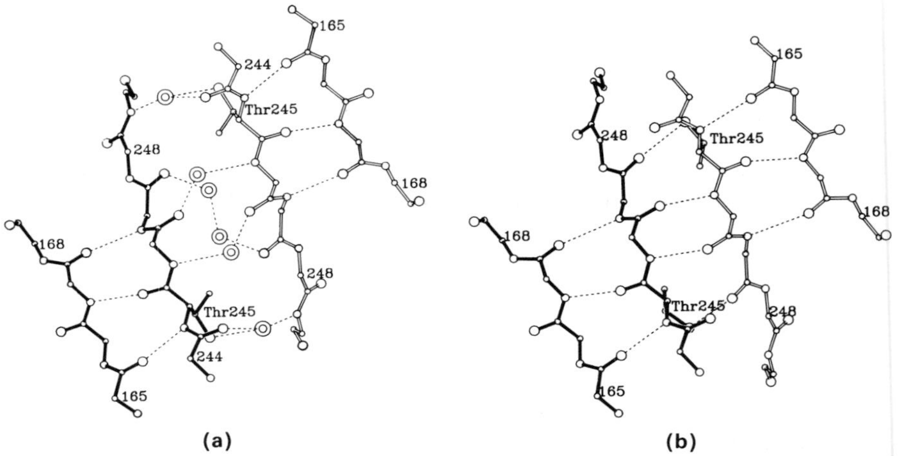

Figure 4.51 Water molecules inserted between strands of β-sheet in allosteric change of phosphofructokinase. (Courtesy of P. R. Evans (92).)

9 Conclusions

In this chapter we have looked at some of the variety in the spatial arrangement of secondary structural elements in globular proteins, and some of the principles governing their mutual spatial relationships. There are those who claim that we have already seen most of the possible structural types, that additional crystal structure determinations will produce only relatives of known structures. However, if one follows closely the new structures reported each year, many are similar to known structures—but there are always a few that show some truly novel feature. (Aficionados distinguish between, 'Oh well . . .' structures and 'Oh, wow!' structures.) The chances are good that there is a lot more left to be discovered. In writing these pages, I wonder what it would be like to write such an essay in—what? 30 years?—when we really do know the structure of every protein of *E. coli* (at least). Presumably it will be possible to give a fairly complete account of what types of proteins exist. This should compensate for the loss of the excitement we now have in *not* knowing what unexpected new structures the next few months will bring.

5

Structure comparison and structural change

1 Introduction

In a room at the MRC Laboratory of Molecular Biology in Cambridge, physical skeletal models of the structures of oxy and deoxy haemoglobin, constructed of brass wire components, stand side by side. At a scale of 2 cm per ångstrom, each is about 1.5 metres in diameter. For many purposes, these are still better than computer graphics devices: they contain fully detailed structural information (except for a few bits that have fallen away). All the geometric transformations necessary to create an image are done automatically. (To move closer to some region, you do not have to hunt in the dark for a dial.) One need not even 'book time' to look at them.

But ... how does one identify the conformational changes involved in the allosteric transition? How does one describe the structural differences between α and β chains that the many amino acid sequence changes have produced?

There are no satisfactory answers to these questions using physical models. Using computer graphics, however, it is possible to superpose two (or more) structures in one picture, and this can reveal—immediately and obviously—which features are conserved and which are changing. It is this facility, more than the 'logistical' ones of finding space for many physical models and keeping them intact, in which lies the real advantage of computer graphics.

In preceding chapters we have examined the ability of computer graphics to let us look at structures, *one at a time*, in very versatile ways. In this chapter we discuss how to use these methods to study structural changes, and to compare structures.

There are two ways to superpose structures. One may draw pictures of each structure, and use interactive graphics to provide the facility to rotate and translate one with respect to the other, superposing the two 'by eye'. The second method is numerical. By selecting corresponding sets of atoms from two structures, a program can calculate the best 'least-squares fit' of one set of atoms to another. Both methods are useful: Fitting by eye permits rapid experimentation to assess the goodness of fit of different portions of the molecules. Numerical fitting permits quantitative comparisons of the relative goodness of fit of different structures and substructures. It also is a useful adjunct to many graphics-related tasks, such as the search in a database for a fragment that can bridge a gap between two segments of chain (see ref. 85). With the very high level of computational power available in the new graphics workstations, it will be possible to design software that combines the advantages of both approaches.

2 Numerical superpositions

Suppose we are dealing with two or more protein structures that contain regions in which the backbone atoms are almost congruent. We wish to superpose the two structures by moving one with respect to the other so that the corresponding atoms in the well-fitting regions are optimally matched.

There are two aspects to this problem:

1. Choosing the corresponding sets of atoms from the two structures, and
2. Finding the best match of the corresponding atoms.

The first of these problems is neither easy nor straightforward when the sequences cannot be aligned. One approach to solving it is to use the solution to problem 2 to perform many exploratory calculations; let us therefore place problem 1 to one side temporarily.

2.1 Rigid body motions and least-squares fitting

The problem that mathematics can solve directly and easily is this: given two sets of points: x_i, $i=1, \ldots N$ and y_i, $i=1, \ldots N$ (here x_i and y_i stand for vectors specifying atomic coordinates), find the best rigid motion of the points $y_i \rightarrow Y_i$ such that the sum of the squares of the deviations:

$$\sum_i |x_i - Y_i|^2$$

is a minimum.

Two mathematical facts that we state without proof are:

(a) The most general motion of a rigid body is a combination of a rotation and a translation.

(b) At the minimum, the mean positions (colloquially, the centres of gravity) of the two sets of points coincide.

It follows that Y_i must be equal to $Ry_i + t$ where R is a rotation matrix and t is a translation vector, and that to minimize

$$\sum_i |x_i - (Ry_i + t)|^2$$

a program must choose t to move the set y_i so that its centre of gravity coincides with that of the x_i, and determine the optimal rotation matrix R.

There is an extensive literature treating this problem. In the computer science literature, it appears as the 'orthogonal Procrustes problem' for which Golub and Van Loan (93) offer a numerically stable solution (i.e. one that works even in pathological cases) based on the singular value decomposition of the correlation matrix. For our purposes it suffices to point out that there exist fast and safe solutions.

What these solutions provide is the following:

1. The minimum value of the r.m.s. deviation between corresponding atoms:

$$\Delta = \sqrt{\frac{\sum_i [x_i - (Ry_i + t)]^2}{N}}$$

This quantity is a useful measure of how similar the structures are (see *Figure 5.1*).

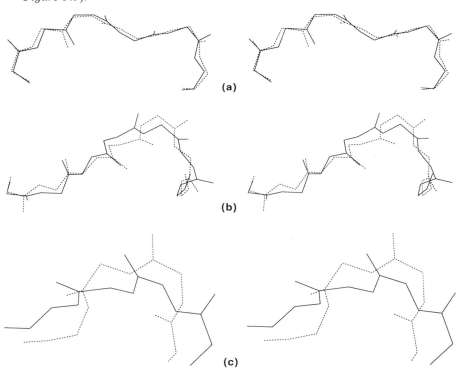

(a)

(b)

(c)

Figure 5.1 What is the significance of different values of the r.m.s. deviation of the two superposed structures? Here are three examples, taken from loops from immunoglobulin structures. (a) Loops L1 (see Section 7.1 in Chapter 4) from J539 [1 FBJ] (solid lines) and HyHEL-5 [2HFL] (broken lines). The r.m.s. deviation of all main-chain atoms (N, Cα, C, O) is 0.51 Å. These structures are the same to within experimental error. (b) Loops H1 from NEWM [3FAB] (solid lines) and McPC603 [1 MCP] (broken lines). The r.m.s. deviation of all mainchain atoms is 1.02 Å. The conformations are quite similar, except that one or two peptides are 'flipped'. (c) Loops H2 from KOL [1 FB4] (solid lines) and HyHEL-5 [2HFL] (broken lines). The r.m.s. deviation of the main-chain atoms is 1.9 Å. The conformations of these loops are quite different. Can the reader now estimate the r.m.s. deviations of the pairs of loops shown in *Figures 4.36b* and *4.38b*?

2. A translation vector t, and rotation matrix R. From the rotation matrix, one can derive the angle of rotation θ from the standard formula $R_{11} + R_{22} + R_{33} = 1 + 2 \cos \theta$, and the direction cosines (l, m, n) of the axis of rotation:

$$(l, m, n) = \frac{-1}{2 \sin \theta} (R_{23} - R_{32}, R_{13} - R_{31}, R_{21} - R_{12}).$$

If $\theta = 0$ the axis of rotation is indeterminate. If $\theta = 180$ degrees, let $v = (R_{11} + 1, R_{21}, R_{31})$; then $(l, m, n) = v/|v|$. Diamond (27) has collected a useful compendium of formulae related to rotation matrices and allied topics.

3. By calculating the transformed points $Y_i = Ry_i + t$, one can report the deviations in *individual* atomic positions $|x_i - Y_i|$. Scanning such a list can reveal that certain portions of the selected regions fit well and that others do not.

Noting which regions fit well and which do not is an important step in the analysis of the structural differences. To refine this comparison, or to work towards an optimal superposition of the well-fitting 'cores' of the structure, a logical next step might be to delete the ill-fitting regions from both lists of atoms, and refit the remainder. (It will be recognized that the first calculation produced a transformation that was a compromise between the well-fitting and ill-fitting portions of the structure, and because of the least-*squares* formulation of the problem the ill-fitting regions will be strongly weighted: Therefore another superposition limited to the well-fitting points should produce a better transformation.)

But this brings us back to the problem of how to choose what atoms to fit.

2.2 Use of least-squares fitting to extract well-fitting substructures: applications to alignment

There are a number of interesting questions that cannot be formulated directly as least-squares fits, as described in the previous section, but which can make use of such calculations. These include:

1. Given a set of corresponding atoms in two proteins, some of which fit well and some of which fit poorly, extract the maximal subset that fits with a specified r.m.s. deviation or better.

2. Given two protein structures, what are the largest well-fitting substructures?

3. Given two protein structures, for which it is not clear how to align the residues on the basis of amino acid sequence alone, can one work out the optimal atom-to-atom correspondence and derive an alignment from the structures?

4. If the 'core' of a pair of protein structures is the union of regions containing corresponding major elements of secondary structure from the two proteins plus flanking peptides, so long as the chains continue to follow the same spatial course, then, given two proteins, determine the residues in the core and the

r.m.s. deviation of their backbone atoms after optimal superposition of the backbone atoms of the entire core together (see refs. 94 and 95 and Section 5.1).

Each of these problems has been approached by application of least-squares fitting calculations.

2.2.1 Extraction of a well-fitting subset

Assume that we have a set of atom–atom correspondences: x_i, y_i, $i = 1, \ldots N$ taken from two proteins for which, for example, we may wish to choose a common orientation. The r.m.s. deviation Δ of all N atoms is larger than we wish, perhaps because of some ill-fitting regions in loops, for example. We can set a threshold, e.g. $\Delta = 0.3$ Å, and a program can seek to reduce Δ to 0.3 Å by carrying out the following steps:

Step 1. Calculate optimal superposition of remaining atoms in the set.

Step 2. Is $\Delta < 0.3$ Å? If so, stop.

Step 3. Find the pair of corresponding atoms that lie the farthest apart after the most recent superposition and delete them from the set. Then, providing any atoms remain, go to step 1.

This procedure is called 'sieving'. This is a fairly simple example, but it illustrates how solutions of complex problems may be based on repeated least-squares fitting of different sets of atoms.

2.2.2 Identification of similar substructures

This and the alignment problem are questions of great interest, but for which no completely satisfactory solution exists. (See, for example, refs 94 and 95.)

One difficulty is that it is hard to formulate the problem in such a way that the answer is unique. For this reason the use of interactive graphics (superposition by eye) as a guide to formulating the appropriate question is strongly recommended. When in doubt look at the structures. (When not in doubt, look at the structures anyway.)

A basic idea is that if some set of corresponding residues from each of two proteins fits well, then any subsets of corresponding residues will also fit well. The converse is not necessarily true: if two well-fitting sets of residues are fitted jointly, the spatial relations between them may be inconsistent and the joint fit may be poor. This suggests that one might start by identifying small well-fitting regions and try to build up larger well-fitting regions by combining them.

It is useful to set a threshold for goodness of fit, for example 0.4 Å. Then one might begin by fitting every N-residue oligopeptide from one structure to every N-residue oligopeptide of the other. Typically, $N \approx 8$. By recording only those pairs of oligomers for which $\Delta \leqslant 0.4$, one has generated a list of pairs of well-fitting substructures. The next step is to work towards larger substructures by merging entries in this list. This procedure can be iterative: given a list of pairs of well-fitting substructures, form the unions of pairs of entries, fit them, and append to the list any new well-fitting substructures discovered. Continue this process until a cycle in which nothing new turns up. Then sort the entries on length and take the top results. Several programs, including WHAT IF and O (see Chap. 6), implement procedures of this general nature and these can be quite useful provided they are used with caution.

Another approach to this problem has been described by Levine, Stuart, and Williams (96). They characterize the conformation by the sequence of main-chain conformational angles φ and ψ, and extract sets of consecutive residues with similar conformations. The advantage of this approach is that it makes it possible to apply the algorithms for alignment of sequences with gaps. In many ways this is a very attractive approach, and it is to be hoped that this work will be developed further.

3 Structural changes arising from changes in state of ligation

There are now numerous cases of proteins for which structures have been determined in more than one state of ligation. In some cases, the structure undergoes little change, except perhaps for specific and localized changes associated with particular functional residues. We saw an example of this in triose phosphate isomerase, *Figure 4.32*. Sperm whale myoglobin is an interesting example in which the oxy and deoxy forms are very similar in conformation; in contrast to haemoglobin, in which ligand binding leads to changes in tertiary and quaternary structure. In other cases, such as insulin, there are specific localized changes associated with the formation of a new, off-axial zinc-binding site. Some enzymes such as citrate synthase show conformational changes in which binding of ligands in a site between two domains leads to a closure of the inter-domain cleft around them (97). Finally, there are the long-range integrated conformational changes associated with allosteric transitions as in haemoglobin, aspartate transcarbamylase, phosphofructokinase, and phosphorylase (98).

3.1 Sperm whale myoglobin

The allosteric change of haemoglobin upon binding oxygen is so famous that many people are surprised to learn that the structures of oxymyoglobin and deoxymyoglobin are extremely similar. The most accurate coordinate sets are based on crystal-structure analyses of sperm whale myoglobin by S. E. V. Phillips (99), who worked with crystals of sperm whale myoglobin in different states of ligation under carefully controlled conditions of solvent and temperature.

The main chains of the oxy and deoxy forms are extremely similar except for three residues at the C-terminus that are somewhat disordered. The optimal r.m.s. deviation of all backbone atoms (N, Cα, C, O) of residues 1–150 is 0.09 Å! (The entire chain is 153 residues long.) Indeed, the optimal r.m.s. deviation of *all* atoms in these residues is 0.23 Å, all of the large differences arising from atoms at the ends of long side-chains. The optimal r.m.s. deviation of all N, Cα, C, O, Cβ, Cγ, and Oγ atoms in residues 1–150 is 0.11 Å.

Figure 5.2a–c shows a superposition of part of oxymyoglobin (solid lines) and deoxymyoglobin (broken lines). This picture includes the haem group, the F-helix and the iron-bound proximal histidine, the distal histidine, and the O_2 of the ligated structure. The lengths of bonds provide a scale: the reader will see that the largest shifts are about a fifth of a bond length, or 0.1–0.2 Å, which is comparable to the experimental error, even in these structures which represent some of the highest-quality protein structure determinations. In fact there are hints of a small concerted shift of the F helix by about 0.1 Å between the oxy- and deoxy-myoglobin structures (C. Chothia, personal communication). Of course, the fact that these shifts are similar in magnitude to the experimental error does not mean that shifts of this size are functionally insignificant! (See the discussion of haemoglobin in Section 3.4.)

This then represents the control: two structures that are virtually the *same* within (the unusually low) experimental error.

It is interesting that, even though the ligated and unligated states are so similar in structure, the molecule must open up during the process of oxygen capture or release: *Figure 5.2d* shows that access to the oxygen-binding site is blocked.

3.2 Insulin: the 'helix interface shear' mechanism of conformational change

Insulin is an interesting system in which to study conformational changes, because it illustrates the effects both of crystal-packing forces and changes in state of ligation.

The pig insulin monomer consists of two chains, A (21 residues) and B (30 residues). The chains are linked by two disulphide bridges; there is a third disulphide bridge within the A-chain. The A-chain contains two helices; the B-chain contains a helix and a strand of sheet (*Figure 5.3a*).

Two monomers form a dimer, held together by hydrogen bonding between strands of β-sheet and by Van der Waals contacts. In the presence of Zn^{2+}, the dimers assemble into hexamers. Crystals grown at low ionic strength contain the 2Zn form, with two zinc ions per hexamer. At high ionic strength an alternative form is produced containing four zinc ions per hexamer. Both the 2Zn and 4Zn forms of pig insulin have been solved to high resolution (see *Figures 5.3* and *5.4*, and *Plates 11 and 12*). The asymmetric unit of each crystal contains two monomers—thus each hexamer can be thought of as a trimer of dimers. The three dimers are related by a crystallographic (i.e. an exact) threefold symmetry and are therefore identical in structure. The two monomers of each dimer are related by a

Figure 5.2 (a–c) Oxy and deoxy forms of sperm whale myoglobin: haem group and F-helix [1MBO, 1MBD]. (d) Blocking of access to oxygen-binding site.

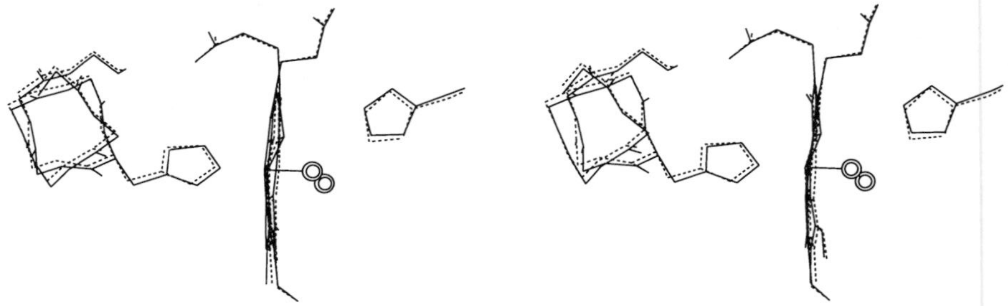

(a)

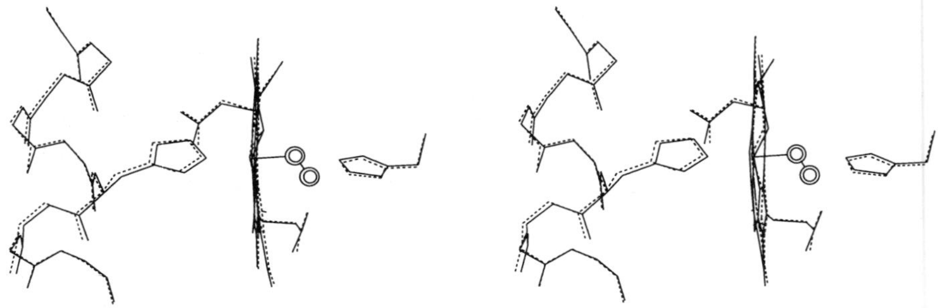

(b)

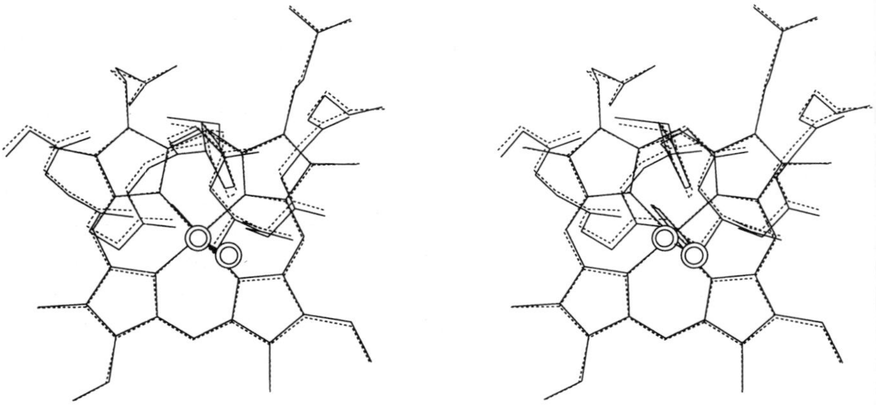

(c)

136

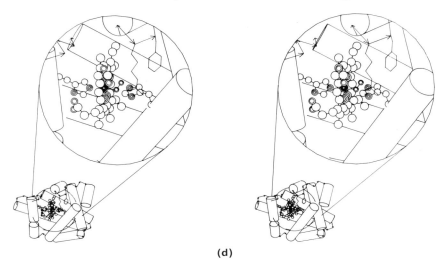

(d)

non-crystallographic, and only approximate, twofold axis. Therefore, in each crystal form, there are two different monomer environments; in each, two different monomeric conformational states are produced.

Symmetry conditions dictate that the two zinc ions in 2Zn insulin must fall on the threefold axis, and that at least one of the four zinc ions in 4Zn insulin must be on the axis. One axial zinc binding site is very similar in 2Zn and 4Zn forms; the 4Zn form contains three off-axial zinc binding sites related by the threefold symmetry.

The 2Zn and 4Zn crystal structures present four independent monomer structures. Of these, one monomer of the 2Zn form is very similar in conformation to the corresponding monomer in the 4Zn form, and can serve as a 'reference structure' for analysis of conformational changes and the search for their origin. In the 2Zn form, the second monomer is quite similar to the reference monomer (*Figure 5a*). In the second monomer of the 4Zn form the N-terminus of the B chain has changed its conformation entirely in the process of forming the off-axial zinc binding site (*Figure 5b* and *Plate 10*).

A detailed comparison of the main-chain conformation of the insulin monomers suggests a physical picture of the deformations: The chains contain many regions (3–10 residues in length) which have main-chain conformations very similar to the same residues in the reference structure. However, these regions of locally-conserved structure have different mutual relative geometries in the reference structure and the second monomers; that is, they are shifted and rotated with respect to one another. The structures can be treated as a succession of nearly-rigid segments of chain connected by more flexible regions. The well-fitting regions can be fit to the homologous regions of the reference molecule with r.m.s. deviations of main-chain atoms of only 0.11–0.25 Å. The only significant segment that exhibits a gross change in topology is the N-terminus of the B-chain,

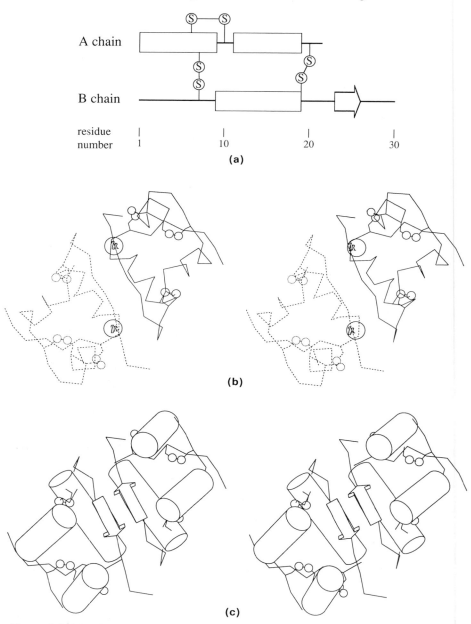

Figure 5.3 (a) The secondary structure and disulphide bridges in the 2Zn insulin monomers. The strand in the B chain forms a β-sheet with the corresponding strand in another monomer. (b, c) The 2Zn insulin dimer [1INS]. (b) Chain trace. (c) Schematic representation. The threefold axis relating this dimer to the other two dimers in the hexamer is vertical in the plane of the page, passing through the two zinc atoms.

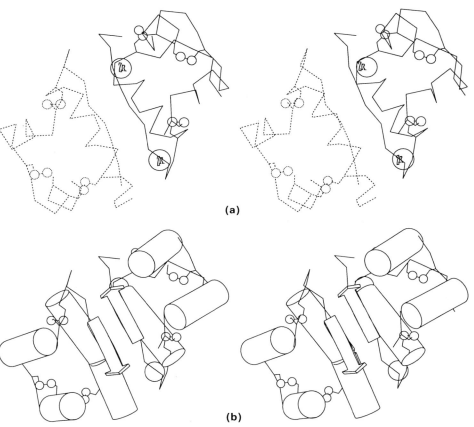

Figure 5.4 The 4Zn insulin dimer; in same orientation as the 2Zn dimer in *Figure 5.3*. (a) Chain trace; (b) schematic representation. (Courtesy of G. G. Dodson.)

in one of the molecules of the 4Zn form; this change is associated with the formation of the off-axial zinc binding site.

These nearly-rigid segments have moved with respect to one another. The magnitudes of the movements are in most cases no more than about 1.5 Å. How are these movements accommodated: why does the close-packing of interfaces between packed helices not hinder them severely? The shifts are made possible by small changes in conformational angles which allow the side-chains in the interfaces to shift. These changes in conformation are usually quite small; the torsion angles only rarely flip from one local minimum to another. (The energy surface observed in twisting a conformational angle typically contains local minima, arising from barriers to internal rotation around single bonds created by the higher energy of eclipsed relative to staggered conformations. Interactions with other atoms of the protein tend to sharpen and accentuate these topographic features of the conformational energy surface.) It is this condition of being trapped in local minima that seems to impose a limit on the magnitude of the shift of one helix relative to another.

The conformational change in insulin occurs as follows: shifts in packed helices are facilitated by small conformational changes within helix interfaces, which permit finite but limited displacements of the main-chain atoms. This 'helix interface shear' mechanism of conformational change allows a maximum displacement of packed helices of about 1.5 Å, which appears to represent the limit of plastic deformation of the helix interface. These shifts can be seen in *Figure 5.5*, using the 1.5 Å rise residue in an α-helix as a measure of distance.

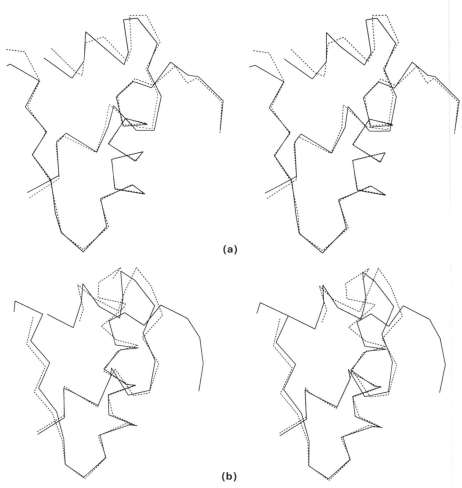

(a)

(b)

Figure 5.5 Comparison of main-chain conformation of the two monomers of (a) 2Zn insulin, (b) 4Zn insulin.

This mechanism has interesting general implications for the long-range transmission of conformational change. Consider two idealized extremes: If proteins were infinitely 'soft', any local conformational change would be dissipated in the immediate vicinity of the perturbation, and could not be transmitted further. If proteins were infinitely 'hard', any local conformational change could cause only global movements of a rigid unit, and not allow deformations of the type required by 'induced fit' of enzymes to substrates, or allosteric changes in tertiary structure. The observation that proteins do have a potential for deformation, but a limited one, shows how conformational change can be transmitted over long distances, and even amplified.

These observations explain an old puzzle about the insulin structures: the reason for the difference in conformation of Phe B25 between the two monomers in both 2Zn and 4Zn insulin. (See *Figure 5.6*.) In one molecule (the reference structure) the ring of Phe B25 packs into its own monomer; in the other the ring points out across the dimer interface to pack against the other monomer. This change takes place because the pocket occupied by Phe B25 in the reference structure is deformed in each of the other monomers. This deformation is the result of the transmission of conformational change from a perturbation arising from crystal packing forces, initiated at a site 20 Å away at the surface of the hexamer. (See ref. 100 for details).

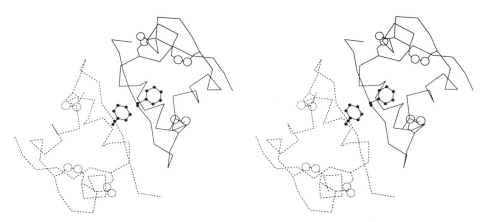

Figure 5.6 Positions of Phe B25 in the two monomers of 2Zn insulin [1INS].

3.3 Citrate synthase

Many enzymes change conformation in response to the binding of substrates and cofactors. Often the active site occupies a cleft between two domains, and binding of co-factors is accompanied by the closing of the cleft over the ligands. Functional reasons for such a conformational change may include the necessity to orient catalytic groups around the substrate, discrimination against unwanted competitive ligands, and the exclusion of water from the active site. These conformational changes occur in some proteins that show no allosteric behaviour.

An example of such closure of an interdomain cleft occurs in the enzyme citrate synthase, a large dimeric protein (97). The crystal structures of two forms of citrate synthase were determined by Remington, Wiegand, and Huber (101). The monomer (of about 540 residues) contains 20 helices that form a structure containing two domains of unequal size (see *Figure 5.7a*). One crystal structure contains the substrate but not the cofactor coenzyme A; in the other the protein binds the substrate, citrate, and coenzyme A (*Figure 5.7b*). In the unliganded

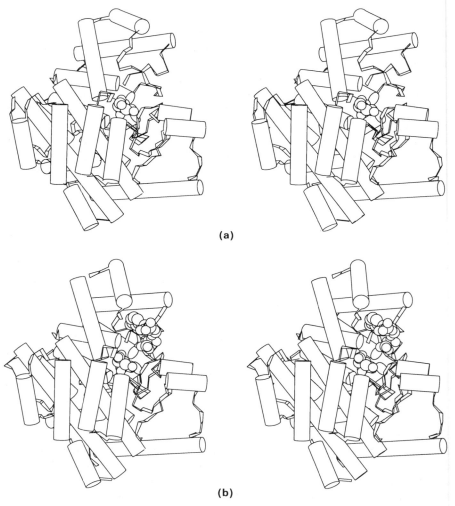

(a)

(b)

Figure 5.7 (a) Open and (b) closed forms of citrate synthase [3CTS, 5CTS].

form the cleft between the two domains is open; in the ligated form the molecule has changed conformation so as to bury almost completely the substrate and co-factor. A loop between two helices (O and P) moves by 6 Å and rotates by 28 degrees to cover the ligands and form hydrogen bonds to them. Some atoms move as much as 10 Å.

Citrate synthase has an extensive inter-domain interface, which is incompatible with purely 'rigid-body' movements of the domains. Indeed, in citrate synthase there is considerable conformational change within the domains, which contributes to the observed closure of the cleft. Seven helices of the large domain together have a very similar structure in both forms of the molecule. These form a semi-rigid kernel, which one may note contains only about one-third of the helices and is limited to one of the domains. The other thirteen helices reorganize their relative spatial disposition, both with respect to the set of seven and in most cases with respect to one another. At the domain interface, the relative movements of packed helices vary from 0.2 Å displacement and 4 degrees of rotation, up to 1.8 Å displacement and 11 degrees of rotation.

How then do the large motions required to close the inter-domain cleft occur? Within the limits of the data, the helix movements are consistent with the 'helix interface shear' mechanism observed in insulin. The individual relative displacements of pairs of packed helices are coupled to produce the large shifts as a cumulative effect. Because there is a limit to the excursion of any single pair of packed helices, the large conformational changes must be built up from the cumulative effects of small ones.

3.4 Haemoglobin

To play its physiological role in oxygen distribution effectively, haemoglobin must capture oxygen in the lungs as efficiently as possible, and release as much as possible to other tissues. To achieve this 'take from the rich, give to the poor' effect, haemoglobin has a high oxygen affinity at high pO_2 and a low affinity at low pO_2. The binding is said to be co-operative, in that binding of some oxygen enhances binding of additional oxygen.

The vertebrate haemoglobin molecule is a tetramer containing two α chains and two β chains. In normal adult human haemoglobin, each α chain has 141 amino acids and each β chain has 146. The α and β chains both resemble their monomeric homologue myoglobin in amino acid sequence and in three-dimensional structure (see *Figure 1.3*). These (and other) globins diverged from each other after gene duplications, myoglobin splitting from haemoglobin about 600–800 million years ago and the α and β chains of haemoglobin splitting about 500 million years ago (see ref. 51).

In vivo, the haemoglobin tetramer can have either of two structural forms. One is characteristic of deoxy (unligated) haemoglobin, and has low oxygen affinity; the other is characteristic of oxyhaemoglobin (four oxygen molecules bound) and has high oxygen affinity. In the erythrocyte, haemoglobin is an equilibrium mixture of the two forms; the concentration of partially-ligated forms is tiny.

(However, the structures of partially-ligated haemoglobin can be observed in crystal structures; see ref. 102). Binding of oxygen induces structural changes that alter the relative free energies of the two forms, shifting the equilibrium towards the high-affinity form. Starting from the deoxy structure, partial ligation (between two and three oxygens) is enough to shift the equilibrium in favour of the other state, which will pick up the remaining oxygens with greater affinity. Conversely, starting with the fully-ligated oxy structure, loss of one to two oxygens will shift the equilibrium to the low-affinity state, stimulating the release of the remaining oxygen. Other molecules that modify oxygen affinity, such as diphosphoglyceric acid, a natural allosteric effector, operate in part by shifting this equilibrium by preferentially stabilizing one of the two forms.

The oxygen affinity of the oxy form of haemoglobin is similar in magnitude to that of isolated α and β subunits, and to that of myoglobin. The oxygen affinity of the deoxy form is much less: The ratio of the binding constants for the first and fourth oxygens is $1:150$–300, depending on the conditions. Therefore it is the deoxy form that is special, which has had its oxygen affinity 'artificially' reduced. (It would thus be more correct to speak of a co-operative *release* of oxygen than a co-operative *binding* of oxygen.) In the terminology of Monod, Wyman and Changeux (103), in their general theory of allosteric change, the reduced oxygen affinity of the deoxy form of haemoglobin is achieved by structural constraints that place the structure in a 'tense' (T), state; while the oxy form in a 'relaxed' (R) state is as free to bind oxygen as the isolated monomer. In fact, the 'tense' state is strained only when it binds oxygen; to call the unligated deoxy state 'tense' is somewhat misleading.

The binding of oxygen is accompanied by a change in the state of the iron. In the deoxy state, the iron has five ligands—the four pyrrole nitrogens of the haem group and the proximal histidine—and is in a high-spin Fe (II) state with an ionic radius of 2.06 Å. In the oxy state, the iron has six ligands—oxygen being the sixth—and is in a low-spin Fe (II) state, with an ionic radius of 1.98 Å. The radii are important because the distance from the pyrrole nitrogens to the centre of the haem is 2.03 Å; this implies that the iron will fit in the plane of the pyrrole nitrogens in the oxy state but not in the deoxy state.

Table 5.1 summarizes the nomenclature and properties of oxy and deoxy haemoglobin.

3.5 Structural differences between deoxy and oxy haemoglobin

The structures of haemoglobin in different states of ligation have been studied with intense interest, partly because of their physiological and medical importance, and partly because they were thought to offer a paradigm of the mechanism of allosteric change (104, 105). (As the structures of other proteins showing allosteric changes have been determined, it is becoming apparent that

Table 5.1 Oxy and deoxy forms of human haemoglobin

	Oxy	**Deoxy**
Oxygen	High ($K_4 = 0.17$ mm Hg)	Low ($K_1 = 26$ mm Hg)
Spin state of iron	low spin	high spin
$r(Fe^{2+})$	1.98 Å	2.06 Å
Monod, Wyman, and Changeux interpretation	tense (T) state	relaxed (R) state
Structure determination: Resolution	2.1 Å	1.74 Å
Reference	Shaanan (105)	Fermi *et al.* (106)

different systems achieve co-operativity by somewhat different mechanisms; see ref. 98.)

The two crucial questions to ask of the haemoglobin structures are;

1. What is the mechanism by which the oxygen affinity of the deoxy form is reduced?

2. How is the equilibrium between low- and high-affinity states altered by oxygen binding and release?

Comparison of the oxy and deoxy structures has defined the changes in tertiary structures of individual subunits; and in the quaternary structure, or the relative geometry of the subunits and the interactions at their interfaces (106). These changes are coupled: as a preliminary and purely qualitative description we point out that the quaternary structure is determined by the way the subunits fit together, which depends on the shape of their surface. The tertiary structural changes alter the shapes of these surfaces, changing the way they fit together.

3.5.1 Tertiary structural changes between deoxy and oxy haemoglobin

The tertiary structural changes in α and β subunits are similar but not identical.

At the haem group itself, in the deoxy form the iron atom is out of the plane of the four pyrrole nitrogens of the haem group. There are two reasons for this: the larger radius of the iron in its high-spin state, and steric repulsions between the Nε of the proximal histidine and the pyrrole nitrogens (107, 108). The haem group is folded, or 'domed'; that is, the iron-bound nitrogens of the pyrrole rings are out of the plane of the carbon atoms of the porphyrin ring of the haem, by 0.16 Å in the α subunit and 0.10 Å in the β subunit.

Forming the link between iron and oxygen would, in the absence of tertiary structural change, create strain in the structure: without constraint the haem would unfold to become planar, and the iron would move into this plane, but these changes are resisted by the steric interactions between the proximal histidine and the haem group, and the packing of the FG-corner (that is, the region between the C-terminus of the F-helix and the N-terminus of the G-helix)

145

against the haem group. In the β subunit, an additional barrier to the binding of oxygen without tertiary structural change is the position of Val E11 in the region of space to be occupied by the oxygen itself.

In the oxy structure, these impediments are relieved by changes in tertiary structure. The FG corners are pushed out, permitting the F-helix to move to allow the proximal histidine and iron to move. Describing these changes locally, relative to the haem group, there is in both subunits a shift of the F-helix across the haem plane (*Figure 5.8*) by about 1 Å, and a rotation relative to the haem plane. The effect is to permit a reorientation of the proximal histidine so that the iron atom can enter the haem plane. Associated with this shift in the F-helix, there are conformational changes in the FG-corner.

To make the connection between tertiary and quaternary structural changes, we must describe the tertiary structural changes in terms of their effects on the shape of the entire subunit: the purely local description of what happens around the haem group is important to rationalize the energetics of ligation but is the wrong frame of reference to account for the change in quaternary structure.

3.5.2 Inter-subunit interactions in haemoglobin

(This section follows closely the discussion of Baldwin and Chothia (104).)

The haemoglobin tetramer can be thought of as a pair of dimers: $\alpha_1\beta_1$ and $\alpha_2\beta_2$. In the allosteric change, the $\alpha_1\beta_1$ and $\alpha_2\beta_2$ interfaces retain their structure, as does a portion of the molecule adjoining these interfaces, including the B,C,G,H regions of both subunits and the D-helix of the β subunit. The overall allosteric change involves a rotation of 15 degrees of the $\alpha_1\beta_1$ dimer with respect to the $\alpha_2\beta_2$, around an axis approximately perpendicular to their interface. (The motion is like that of a pair of shears with α_1 and $\alpha_2 =$ the blades and β_1 and $\beta_2 =$ the handles. This is illustrated in ref. 104 and in the centrefold accompanying ref. 109.)

Given that the structure of the $\alpha_1\beta_1$ interface is conserved, it provides the appropriate frame of reference for describing the tertiary structural changes in a way that relates them to the changes in surface topography that account for the quaternary structural change.

With respect to the $\alpha_1\beta_1$ interface, the tertiary structural changes appear as follows (*Fig. 5.9a*): in going from deoxy to oxy structures in both subunits, the haem groups move into the haem pockets, ending up 2 Å closer together in oxy than in deoxy structures. The backbone of the F-helix moves with the haem (it *also* moves *relative* to the haem as discussed above). The FG-corner also shifts. These tertiary structural changes of the F-helix and FG-corner do not extend beyond the EF-corner on the N-terminal side and the beginning of the G-helix on the C-terminal side. Note that this is consistent with the statement that the C (and, in β, D) and G regions are in the part of the structure that is conserved in the allosteric change.

To understand the quaternary structural change, we must analyse how the interface between the $\alpha_1\beta_1$ and $\alpha_2\beta_2$ dimers changes. The most important

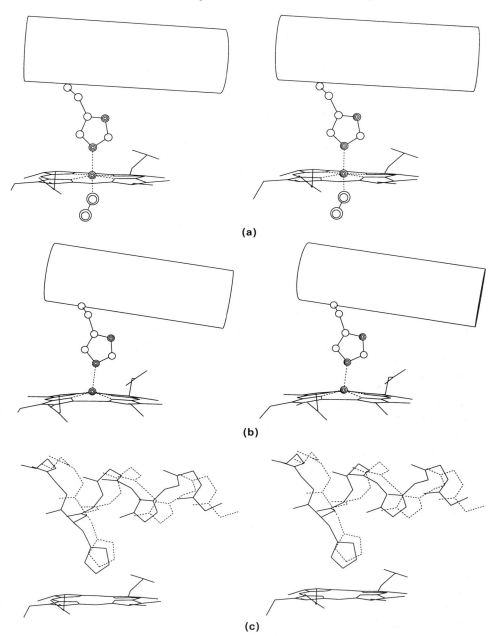

Figure 5.8 Shifts in human haemoglobin as a result in change in state of ligation. This figure shows the F-helix, proximal histidine, and haem group of the *a*-chain in the (a) deoxy and (b) oxy forms of human haemoglobin [1HHO, 4HHB]. (c) Superposition: oxy (solid lines); deoxy (broken lines); only the oxy haem is shown. The structures were superposed on the haem group.

147

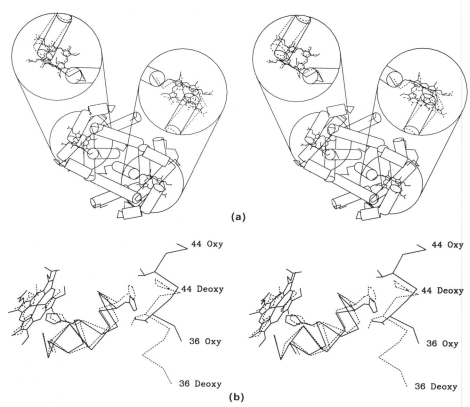

Figure 5.9 Some important structural differences between oxy and deoxy haemoglobin [1HHO, 4HHB]. (a) The $a_1\beta_1$ dimer in oxy (solid lines) and deoxy (broken lines, in 'blown-up' regions only) forms. In the blown up regions only the F-helix, FG-corner, G-helix and haem group are shown. The oxy and deoxy $a_1\beta_1$ dimers have been superposed on their interface; in this frame of reference there is a small shift in the haem groups, and a shift and conformational change in the FG-corners. (b) Alternative packings of a_1 and β_2 subunits in oxy (solid lines) and deoxy haemoglobin (broken lines). Here the oxy and deoxy structures have been superposed on the F- and G-helices of the a_1 monomer. Although for purposes of this illustration we have regarded the a_1, subunit as fixed and the β_2 subunit as mobile, in fact is it only the relative motion that has real significance.

inter-subunit contacts are between the α_1-β_2 and α_2-β_1 subunits. (In the open-shears image, the important variable contacts are between each blade and the opposite handle. The contacts between each blade and its own handle are—in haemoglobin as well as in shears—rigid.)

The interacting regions are α_1 FG-corner—β_2 C-helix and β_2 FG-corner—α_1 C-helix. These are not identical. (However, these interactions are the same by symmetry as α_2 FG-corner—β_1 C-helix and β_1 FG-corner—α_2 C-helix respectively.)

The α_1 FG–β_2 C interaction is very similar in oxy and deoxy structures.

Residues Arg 92 FG4, Asp 94 G1, and Pro 95 G2 of the α_1 subunit are in contact with Tyr 37 C3 and Arg 40 C6 in the β_2 subunit. In comparing oxy and deoxy structures there are small conformational changes in these residues but the pattern of interactions is retained.

The other region of contact, β_2 FG-corner–α_1 C-helix, differs substantially between oxy and deoxy structures (*Figure 5.9b*). In the deoxy structure, His β_2 97 (FG4) packs between Thr α_1 41 (C6) and Pro α_1 (44) (CD2), and there is a hydrogen bond between the side-chains of Asp β_2 99 (G1) and Tyr α_1 42 (C7). In the oxy structure, His β_2 97 packs between Thr α_1 38 (C3) and Thr α_1 41 (C6). (Because the C-helix is a 3_{10} helix, this corresponds to a jump of one turn relative to the His β_2 97 against which it packs. The shift in β_2 FG relative to α_1 C is approximately 6 Å.) The Asp–Tyr hydrogen bond is not made in the oxy structure.

This explains the two discrete quaternary states: The β_2 FG-corner–α_1 C-helix contact has two possible states, depending on the subunit shape presented by the tertiary structural state. The other contact, α_1 FG–β_2 C, changes only slightly.

The requirement for the quaternary structural change arises from the tertiary structural changes; in particular, from the shifts that bring the FG-corners of the α_1 and β_1 subunits in the $\alpha_1\beta_1$ dimer 2.5 Å closer together in the oxy structure than in the deoxy structure. One tertiary structural state of the $\alpha_1\beta_1$ and $\alpha_2\beta_2$ dimers can form a tetramer with one state of packing at the interface; the other tertiary structural state is compatible with the alternative packing.

In conclusion, it may be useful to trace the logical connection between the structural changes: Starting from the deoxy structure, ligation of oxygen creates strain by preventing the haem group from adopting the geometry optimal with the oxygen bound. Relief of strain around the haem group requires moving the F-helix and the FG-corner. To accommodate these changes there must be a set of *tertiary* structural changes which change the overall shape of the $\alpha_1\beta_1$ and $\alpha_2\beta_2$ dimers; notably the shifting of the FG-corners, which in the deoxy structure prevent the F-helix from moving. In consequence, the deoxy *quaternary* structure is destabilized because the dimers no longer fit together properly (having changed their shape). Adopting the alternative quaternary structure requires the tertiary structural changes to take place even in subunits not yet liganded. As a result of the quaternary structural change, they have been brought to a state of higher oxygen affinity. It is important to emphasize that this is a sequence of steps in a *logical* process and not a description of a temporal pathway of a conformational change.

4 Structural changes attendant upon chain cleavage

A number of proteins undergo conformational changes when a specific peptide bond is broken. Unlike the rather small changes associated with the activation of serine proteases, the serine protease inhibitor α_1-antitrypsin undergoes a very dramatic springing open when it is cleaved (*Figure 5.10*). In the cleaved form, the

Figure 5.10 (a) a_1-antitrypsin, cleaved form [5API]. The broken line connects the atoms that form a peptide bond in the uncleaved form. (b, c) Comparison of open and closed forms of molecules in the serine protease inhibitor (serpin) family. (b) a_1-antitrypsin, cleaved, as in part (a). (c) Ovalbumin, uncleaved. It is extremely likely that the uncleaved form of a_1-antitrypsin has a fold similar to that seen in ovalbumin. Corresponding parts of these molecules are shown in bolder lines. Note that a helix and short strand in ovalbumin (c) correspond to a long strand *inserted into the middle of the large sheet* in cleaved a_1-antitrypsin. This change in the topology of the fold is unusual and dramatic. (Ovalbumin picture courtesy of P. E. Stein and A. G. W. Leslie.)

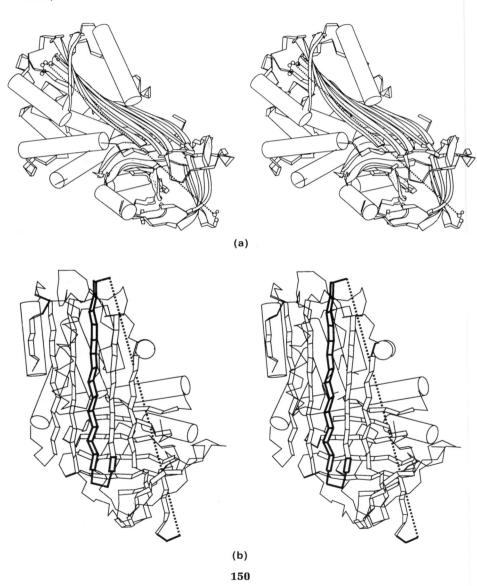

(a)

(b)

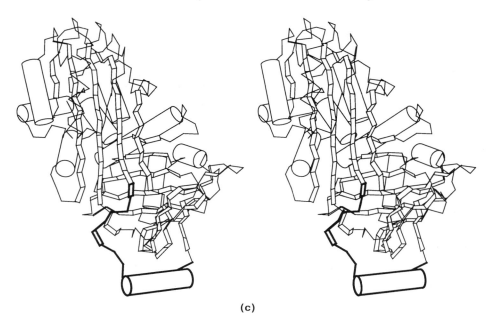

(c)

main-chain C of residue 358 and the N of residue 359 are 65 Å apart! Obviously this molecule must undergo a change in secondary and tertiary structure upon cleavage.

5 Structural relationships among related molecules

Families of related proteins tend to retain similar folding patterns. *Figure 5.11* shows sperm whale myoglobin, its distant relative lupin leghaemoglobin and *Magistocladus laminosus* C-phycocyanin, α-chain. *Figure 1.12* shows two related β-sheet proteins, plastocyanin and azurin. *Figure 5.13* shows the two sulphydryl proteases, actinidin and papain. The reader is urged to spend some time comparing these sets of structures, cataloguing features that remain the same and features that change. Which pair of structures appears to be most closely related?

5.1 A general relationship between divergence of amino acid sequence and protein conformation in families of related proteins

If one examines sets of related proteins (see *Figures 5.11–5.13*) it is clear that the general folding pattern is preserved, but that there are distortions which increase in magnitude as the amino acid sequences diverge. A closer look reveals that the distortion is not uniformly distributed, but that, in any family, there is a core of the structure that retains the same qualitative fold, and other parts of the structure that change conformation radically. An analogy: if one were to compare the capital letters R and B, the common core would correspond to those

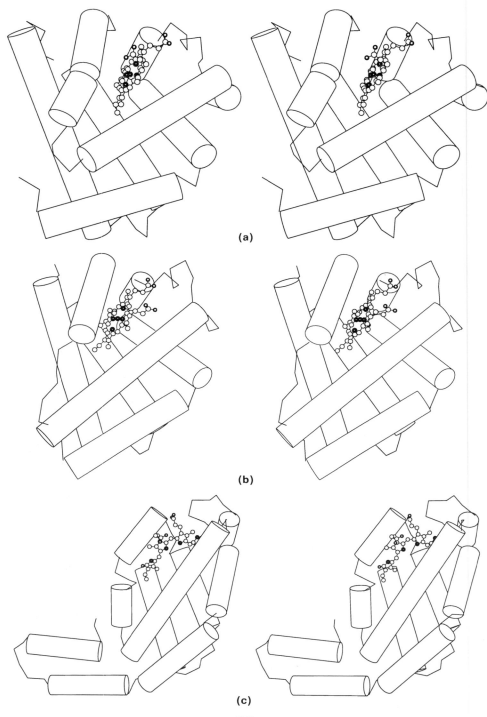

(a)

(b)

(c)

152

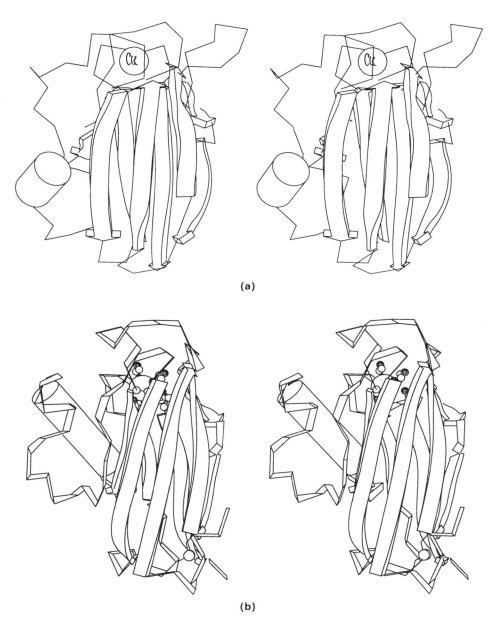

(a)

(b)

Figure 5.12 (a) Poplar leaf plastocyanin [1PCY], and (b) *A. denitrificans* azurin [2AZA], electron transport proteins in green plant photosynthesis and bacterial respiration, respectively.

Figure 5.11 (a) Sperm whale myoglobin [1MBO], (b) lupin leghaemoglobin [2LH4], and (c) *Mastigocladus laminosus* C-phycocyanin, *α*-chain. (Courtesy of T. Schirmer and R. Huber.)

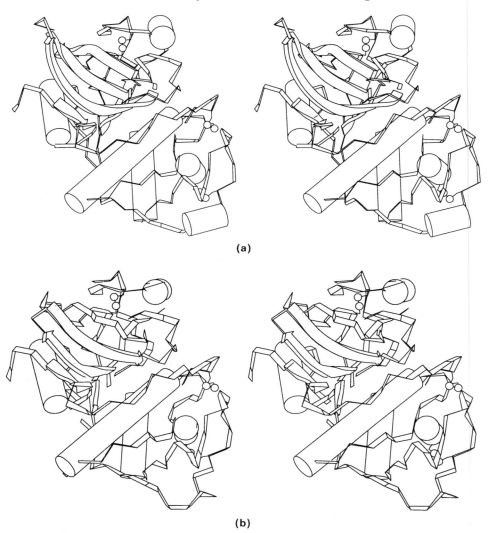

Figure 5.13 (a) Actinidin [2ACT], and (b) papain [8PAP], sulphydryl proteases.

portions of the letters represented by the letter P. Outside this common core, R has a diagonal stroke and B has a loop. Comparison of a roman R and an italic *B* would illustrate a common core that was distorted between the two structures. In proteins, the common core generally contains the major elements of secondary structure and segments flanking them, including active site peptides.

Very large structural changes in the regions outside the core make it difficult to measure structural change quantitatively by straightforward application of simple least-squares superposition techniques. First, some of the region outside the core may have diverged so far in both sequence and structure that it is impossible to align the sequences, to assign corresponding residues to calculate the superposition. Even when it is possible to align regions outside the core, the fact that the superposition procedure minimizes the sum of the *squares* of the deviations renders it overwhelmingly sensitive to regions showing large deviations. As a result, the relatively modest structural changes in the core would be swamped by the large deviations outside it.

To define a useful measure of structural divergence, it is necessary first to extract the core and then carry out the least-squares superposition on the core alone. Lesk and Chothia (110, 111) gave a prescription for doing this: given two related proteins, for each major element of secondary structure perform a succession of superposition calculations, that include the main-chain atoms (N, Cα, C, O) of corresponding secondary structural elements plus additional residues extending from either end. Include more and more additional residues until the α-carbon of the last residue tested deviates by more than 3 Å. One has now identified a well-fitting contiguous region containing an element of secondary structure plus flanking segments. After finding such pieces corresponding to all common major elements of secondary structure, do a joint superposition of the main-chain of all of them. The result of this calculation is the r.m.s. deviation of the core. The major contributions to this quantity are generally the shifts in position and orientation of the regions with respect to one another, not the poorness of fit of the individual regions.

The r.m.s. deviation of the core measures the divergence of the structure. The fraction of identical residues in the core measures the divergence of the sequences.

Figure 5.14 shows the relationship between divergence of sequence and structure for 32 pairs of proteins from 8 different families. It is of the greatest interest and significance that there is a *common* relationship, that holds for all the different families studied.

These results show that as the sequence diverges, the structure also diverges, with an exponential dependence. However, as the sequences diverge, the fraction of residues in the core (the fraction of the structure that retains the same qualitative fold) may also decrease substantially. This is shown in *Figure 5.15*. For sequences more closely related than 50% residue identity, the core is observed to contain at least 95% of the residues, and any refolding of the remaining 5% will involve only minor surface loops. But for very distantly related structures, with residue identities less than 20%, evolution often has had (and has taken) the opportunity to alter far more of the structure—in such cases the core can amount to as little as 40% of the structure or as much as 89–90% (111).

Let us consider three representative points on the curve. First, the points at 100% residue identity represent independent structure determinations of the same protein. The observed differences in structure arise primarily from crystal-packing forces—the molecules have different environments in the crystal and are subjected to different patterns of intermolecular contacts. The average value of

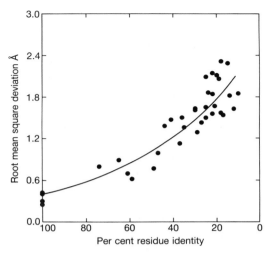

Figure 5.14 Relationship between divergence of sequence and structure in related proteins (From ref. 110.)

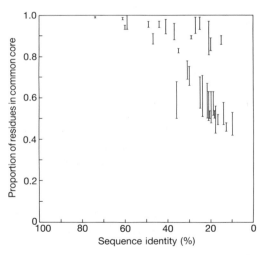

Figure 5.15 Variation in size of core in protein families as a function of sequence divergence (From ref. 110.)

the r.m.s. difference in atomic positions in these pairs of structures is 0.33 Å. The implication is this: protein conformation is determined primarily by amino acid sequence, but modified by subsidiary factors, mainly crystal-packing forces but also the conditions of solvent and temperature. The value 0.33 Å estimates the magnitude of the effects of these secondary factors and, of course, includes a component from experimental error. Because this value is much smaller than the structural differences observed in related but non-identical proteins, the largest part of the structural differences in those cases must genuinely arise from the changes in amino acid sequences.

Consider next the pair of sulphydryl proteases papain and actinidin (*Figure 5.13*). In this case the core contains 206 residues (out of a total of 212 in papain and 218 in actinidin). Thus the core includes almost all of either molecule. The residue identity is 49%, and the r.m.s. deviation of the core is 0.77 Å.

Finally, with less than 20% residue identity, the disturbance to the structure is severe. The core may be limited to no more than 40% of the sequence, with 60% of the chain adopting a quantitatively different fold. The reason for the onset of larger changes below about 50% sequence identity of the core is that in closely related proteins the sequence changes involve primarily surface residues, which exert relatively little leverage on the structural framework. Distantly related proteins show more extensive mutation of buried residues, changes that have more serious effects on the structure of the core.

5.2 Application to model building of homologous proteins: 'Here is a new sequence . . .'—Part II

Suppose that the sequence of an unknown protein is found to be homologous to that of a protein of known structure. Then, the results presented in the preceding section show that the core of the known protein can provide a model for the core of the unknown one, and that the quality of this model can be predicted from the extent of divergence of the sequences.

The level of 50% residue identity—or any closer relationship—provides a useful 'rule of thumb' for the utility of such a model. Thus, suppose one had determined the amino acid sequence of a new sulphydryl protease with 50% residue identity with actinidin, and built a model of this protein by taking the backbone of actinidin and replacing the mutated side-chains, retaining the side-chain conformation of the parent structure whenever possible. If the structure of the new protein were determined by X-ray crystallography, we should expect, on the basis of this relatively close relationship between the amino acid sequences, that:

1. A core of over 90% of the residues would retain a common fold. Fewer than 10% of the residues would be found in loops with radically different conformations.

2. The backbone atoms of the core could be superposed with a r.m.s. deviation of about 1.0 Å or less. The binding site might show even less deviation, as

evolution tends to alter binding sites relatively conservatively, provided that one is dealing with a family of proteins in which function is maintained.

3. The side-chains of 90% of the non-mutated residues, and of 50% of the mutated residues, would have similar conformations.

Such a model would give a reasonable picture of the unknown structure, at a level useful for analysis of its structural and functional properties. If the sequences were more closely related, the quality of the model would be correspondingly improved. Even for molecules for which the sequence identity is no more than 35–40% model building might yield useful results.

For proteins more distantly-related than 35–40% residue identity, the model-building procedure described would be less successful, and one should be discouraged from trying to build a full three-dimensional model. The core might include no more than 40% of the sequence, and the geometry of the core residues would be altered more radically—the r.m.s. deviation would be much higher. The one consolation is that the active site might be relatively well conserved between the known and unknown structures, permitting some analysis of the effects on function of mutations within the active site itself.

Unfortunately, there is no better way of generating a model of an unknown protein from a very distant relative, nor is there any procedure—such as those based on conformational energy calculations or molecular dynamics—that will reliably and significantly improve such a model. This is an active subject of research.

5.3 Point mutations

The smallest unit of sequence change is a single-site mutation. Natural variants show that protein structures can tolerate many **but not all** single-site mutations. Homologous proteins from related species, and polymorphic forms even within single individuals, often differ only at isolated sites. For example, pig insulin, used clinically for the treatment of diabetes in man, differs from human insulin only in one position.

In many cases where crystal structures of closely related proteins are known, the structures are very similar, with only small local deformation apparent around the site of the mutation.

This observation is consistent with our understanding of the mechanism of evolution of naturally-occurring proteins. Evolution is a dynamic process. If one imagines protein sequences as drifting in a multidimensional space (one dimension per residue position), then an isolated point of stability and function, with no stable and functional neighbours, would be unlikely: How would evolution find it? Or if it did arise as a result of some concatenation of very unlikely events—or based on a gene designed and synthesized artificially and released in an organism into the environment—it would be unstable.

Folklore has it that single-site mutations on the surface are usually innocuous;

buried interfaces between elements of secondary structure can tolerate an extra methyl group.

Folklore notwithstanding, many single-site variants are not healthy proteins. Many poorly-functioning mutants show up clinically, of which those of haemoglobin have been most comprehensively and carefully studied. Most involve interior residues, especially those interacting with the haem group. However, the mutation that produces HbS, associated with sickle-cell anemia, produces a correctly-folded protein, which causes clinical problems because its deoxy form aggregates easily.

Of course, natural variants are a subset of all possible variants, which have been subjected to natural selection. Artificial variants produced by site-directed mutagenesis extend our knowledge of what is tolerable. There are a small number of cases of a complete *allumwandlung*: the replacement of a single residue by all 19 others, and the testing of functional properties, *and* the solution of the crystal structures, of all 20 proteins. (The term *allumwandlung* refers to a chess problem in which in different variations a pawn must change upon promotion to all other possible pieces; see ref. 112).

A group at Genentech, working with subtilisin, did such a thorough study of Met 222 and Gly 166 (113). The motivation was that oxidation of Met 222 inactivates the enzyme, and an active enzyme without this problem would be a superior ingredient in laundry detergent.

Matsumura *et al.* (114) have carried out a study of replacements of Ile 3 in bacteriophage T4 lysozyme. *Figure 5.16* shows the effect of the change Ile 3→Tyr. There is in this case only a local deformation of the structures.

Is it possible to predict the structures of single-site mutants, knowing the parent structure? A preliminary question is whether it is possible to decide whether the mutation will cause only a small structural deformation, or more drastic changes. This cannot be done entirely reliably *a priori*, but often it is a justifiable assumption, if it is known experimentally that the mutant is functional.

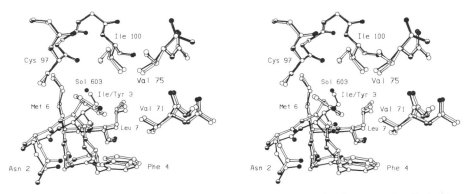

Figure 5.16 Structural change in bacteriophage T4 lysozyme arising from mutation Ile 3→Tyr [3LZM and 1L18]. (Courtesy of B. W. Matthews; from ref. 114.)

The idea would be to use the parent structure, with one residue replaced, as an initial condition for energy minimization or molecular dynamics. But the energy models that underly those calculations are not sufficiently refined: if native structures, with no mutations, are minimized, they suffer distortions which can be of the same order of magnitude or larger than the differences between native and mutant structures. If the calculations are artificially restrained—by tethering the atoms to their initial positions—one risks not allowing genuine changes to happen. It is therefore difficult to predict the small structural changes: one is trying to draw a fine line with a thick brush. This is certainly a problem on which progress would be most welcome.

6

Program systems

In previous chapters I have attempted to show how to look at proteins, necessarily providing only still pictures and primarily monochrome ones. I have, though, emphasized that the most advanced program systems use the most powerful graphics devices available, which have facilities for colour, real-time rotation, depth cueing, and interactive control over the contents and orientation of a picture.

Such devices are now fairly widespread, and it is likely that most readers will find that there exists at their institution something that they can use.

It remains to address the readers who face the question: I want to establish a modern molecular graphics facility; what should I go out and buy? This question is difficult to answer because both the range of hardware available and the program systems for them are changing very rapidly. In many cases, partnerships have been established between hardware manufacturers and software houses.

This chapter deals with a few of the most important programs, and suggests what the prospective purchaser should look for.

1 FRODO

By far the most widely used molecular graphics program is FRODO, written originally by T. Alwyn Jones but subsequently modified and reimplemented for a variety of graphics devices (115). It would be dishonest not to point out that Dr Jones is the leader in the field of molecular graphics—by many lengths—and that he has continued to make outstanding contributions to the field, especially in the development of techniques to support the solution of protein structures by X-ray crystallography. However, many of the methods he has developed are also useful for structural analysis. Other programmers in the field implement his ideas in their own programs—everyone else seems to be struggling to 'keep up with the Jones'.

If FRODO did not exist, it would be necessary to invent it. FRODO was originally designed to assist the protein crystallographer to build a model of a protein into the electron density produced from experimental data. The electron density is usually represented by a set of three-dimensional contours. The protein is represented by a skeletal model, although most versions of FRODO can show dot-surfaces as well. The user has great flexibility in selecting the region in the electron density map to work on, and to build, translate, rotate, and alter the conformation of the model of the protein in that region. There is the opportunity for intelligent division of labour between crystallographer and program: once the human has brought the model to an approximately-correct position and orientation, the program will do a least-squares fit to find the best match in the vicinity.

Versions of FRODO are available for Evans & Sutherland equipment (MPS, PS-300-330-350-390-ESV workstation) and the Silicon Graphics IRIS series.

Two recent advances that Jones and his co-workers have introduced are: (1) the representation of electron densities by their 'skeletons', and (2) the search in a database of known protein structures for pieces to 'paste into' a portion of a structure (85).

1. The 'skeletonization' of a three-dimensional object is a procedure that shrinks the object to a set of lines, while retaining its connectivity. Applied to electron density maps, it produces, in favourable cases, a structure not dissimilar in appearance to the model one wants to build. Starting with an electron-density map of reasonable quality, it is possible to see elements of secondary structure in the skeleton. Knowing the general structural patterns seen in protein structures, the crystallographer can interactively manipulate the connectivity of the skeleton, to extract approximate Cα positions to define the trace of the chain. It would be incorrect to say that the skeletonizing procedure solves the problem (of extracting the chain trace from the electron density); rather, it presents the problem to the crystallographer in a perspicuous way, facilitating the application of *human* wisdom. (Skeletonizing was first applied to this problem by J. Greer, but it seems to require interactive graphics to make it work: the problems of ambiguity at branch points—which branch is main chain, which branch is side-chain, which branch is noise?—are so severe that a human observer is required to resolve them.)

2. Once approximate positions of the Cα atoms have been determined from the skeleton, it is then possible to 'flesh out' the model by appealing to the database of solved protein structures (85). One can choose a set of five consecutive Cα atoms in the skeleton as a search pattern, and look in a database of well-refined structures for a fragment containing five Cα atoms that match the search pattern geometrically. Having found one, or more, such matches, one can simply import a fragment from the database into the model being built. The fragment usually provides good main-chain coordinates. A complete main-chain can generally be built up by repeated application of this procedure to a succession of fragments along the chain.

It is even possible to bridge gaps in the chain by this procedure. One can search in the data base for fragments that fit the Cα atoms that flank the gap in the skeleton, with a specified number of residues intervening. One then imports the main chain of an entire fragment from the database. A particular example of this procedure is the modelling of loops between elements of secondary structure, by searching for fragments of specified length that fit the end of the helices or sheets to be joined.

In most cases, there is no relation between the sequence of the fragment in the skeleton and that of the fragment in the database that one is importing, so that side-chain conformations must be built by fitting to the electron-density map.

The effectiveness of this procedure has revealed that proteins can be considered, at least to a first approximation, as constructed from a relatively small number of 'fixed' fragments of main-chain: Jones has suggested the analogy to Lego. There have been a number of attempts to define the pieces by taking all the oligopeptides from all well-refined structures and clustering them according to similarity of main-chain conformation (116, 117).

2 Other facilities

A number of individuals, many working with commercial software houses, have developed program systems for the display and manipulation of macromolecules. These are generally coupled to an energy minimization and molecular dynamics package, and integrated to greater or lesser extents with databases. All are of very high quality. The most widely used of these include:

Author or company	*Program*
Biosym Technologies	INSIGHT (118)
	DISCOVER
Polygen	QUANTA
BioDesign	BIOGRAF
University of California	
Computer Graphics Laboratory	MIDAS (119, 120)
Tripos	SYBYL, MENDYL
G. Vriend	WHAT IF (121)
T. A. Jones, M. Bergdoll,	O (122)
M. Kjeldgaard	

3 Molecular graphics on a personal computer

The most powerful molecular graphics programs naturally make use of the most powerful equipment: equipment that usually must be shared by a department. However, the power of personal, desk-top computers has increased so much that there is no reason why a fairly powerful graphics system should not be available on an individual PC.

Of particular significance is the project 'Molecular Structures in Biology', a collaboration between the IBM UK Scientific Centre (in Winchester, England), the Protein Data Bank, and Oxford University Press. This system is a personal workstation based on an IBM PS/2, providing access to a database stored on CD-ROM. Facilities include an atlas of protein structures integrated with textual information.

Afterword

A criticism that can be levelled at this book is that it is compendium rather than a synthesis. The treatment of the material is, by and large, descriptive. The problem is that, with the current rate of increase of solved structures, any purely descriptive approach will soon collapse under its own weight.

It is not as if general principles do not exist or are not known. There are empirical generalizations about protein folding patterns, such as the preferred direction of the twist of sheets, or the observation that β–α–β connections have a preferred hand; and there are explanations of some of these rules (see refs 36 and 123). More fundamentally, there are all the well-established principles of physics and chemistry that govern the structure and function of proteins, and even provide clear and simple explanations of some of their features.

But mysteries remain. These include specific mysteries: why I cannot predict, *a priori*, the native structure of sperm whale myoglobin from its amino acid sequence—and general mysteries: how diverse is the full set of spatial patterns that protein chains can describe? Personally, I find the general mysteries more frustrating.

Dirac once wrote:

> ... the main object of physical science is not the provision of pictures, but is the formulation of laws governing phenomena and the application of these laws to the discovery of new phenomena ... One may, however, extend the meaning of the word 'picture' to include any way of looking at the fundamental laws which makes their self-consistency obvious ... (124).

Substituting *for* for *at* in the last sentence gives a fairly precise sense of what we are currently trying to do. This is really the beginning, not the end.

References

1. Schulz, G. E. and Schirmer, R. H. (1979). *Principles of Protein Structure.* Springer-Verlag, Berlin and New York.

2. Creighton, T. E. (1992). *Proteins: Structures and Molecular Properties* (2nd edn). Freeman, New York.

3. Fersht, A. (1985). *Enzyme Structure and Mechanism* (2nd edn). Freeman, New York.

4. Brändén, C.-I. and Tooze, J. (1991). *Protein Structure for Biologists.* Garland Publishing, New York.

5. Rogers, D. F. and Adams, J. A. (1990). *Mathematical Elements for Computer Graphics* (2nd edn). McGraw-Hill, New York.

6. Newman, W. M. and Sproull, R. F. (1979). *Principles of Interactive Computer Graphics* (2nd edn). McGraw-Hill, New York.

7. Angell, I. O. and Griffith, G. (1987). *High-resolution Computer Graphics Using FORTRAN.* Macmillan, London.

8. Foley, J. D., Van Dam, A., Feiner, S., and Hughes, J. (1990). *Computer Graphics: Principles and Practice.* Addison-Wesley, Reading, Mass.

9. Rogers, D. F. (1985). *Procedural Elements for Computer Graphics.* McGraw-Hill, New York.

10. Burger, P. and Gillies, D. (1989). *Interactive Computer Graphics: Functional, Procedural and Device-Level Methods.* Addison-Wesley, Reading, Mass.

11. Lesk, A. M. (1972). Pictorial pattern recognition and the phase problem of X-ray crystallography. *Communications of the Association for Computing Machinery,* **15**, 3–6.

12. Johnson, C. K. (1970). Drawing crystal structures by computer. In *Crystallographic Computing* (ed. F. R. Ahmed, S. R. Hall, and C. Huber), pp. 227–30. Munksgaard, Copenhagen.

13. Lesk, A. M. (1981). *Introduction to Physical Chemistry,* Chapter 14. Prentice-Hall, Englewood Cliffs, N.J.

14. Brünger, A., Kuriyan, J., and Karplus, M. (1987) Crystallographic *R* factor refinement by molecular dynamics. *Science,* **235**, 458–60.

15. Dodson, G. (1986). Protein crystallography and its new revolution. *Trends in Biochemical Sciences,* **11**, 309–10.

16. Brändén, C.-I. and Jones, T. A. (1990). Between objectivity and subjectivity. *Nature,* **343**, 687–9.

17. Abola, E. E., Bernstein, F. C., and Koetzle, T. F. (1988). The Protein Data

Bank. In *Computational Molecular Biology: Sources and Methods for Sequence Analysis* (ed. A. M. Lesk), Ch. 7. Oxford University Press, Oxford.

18. Lesk, A. M. (1983). A toolkit for computational molecular biology. I. Packing and unpacking of protein coordinate sets. *Journal of Molecular Graphics*, **1**, 118–21.

19. Lesk, A. M. (1987). A toolkit for computational molecular biology. III. Micryfon—A (fairly) general program for input of protein coordinate files. *Journal of Applied Crystallography*, **20**, 488–90.

20. Johnson, C. K. (1965). *ORTEP: A Fortran Thermal-Ellipsoid Plot Program for Crystal Structure Illustrations*. ORNL-3794 Revised, Oak Ridge National Laboratory, Oak Ridge, Tennessee.

21. Chothia, C. (1973). Conformation of twisted β-pleated sheets in proteins. *Journal of Molecular Biology*, **75**, 295–302.

22. Adobe Systems, Inc. (1985). *PostScript Language Reference Manual*. Addison-Wesley, Reading, Mass.

23. Adobe Systems, Inc. (1985). *PostScript Language Tutorial and Cookbook*. Addison-Wesley, Reading, Mass.

24. Adobe Systems, Inc. (1988). *PostScript Language Program Design*. Addison-Wesley, Reading, Mass.

25. Brown, E. (1990). A language in control. *Personal Computer World*, **13** (1), 194–6, 198; Smooth operator. *Personal Computer World*, **13** (1), 198–200, 202.

26. Carlbom, I. and Paciorek, J. (1978). Planar geometry projections and viewing transformations. *Computing Surveys*, **10**, 465–502.

27. Diamond, R. (1991). Molecular modeling and graphics. In *International Tables for Crystallography*, Vol. B, Chapter 3.3. Published for the International Union of Crystallography by Kluwer Academic Publishers, Dordrecht.

28. Connolly, M. L. (1983). Solvent-accessible surfaces of proteins and nucleic acids. *Science*, **221**, 709–13.

29. Connolly, M. L. (1983). Analytical molecular surface calculation. *Journal of Applied Crystallography*, **16**, 548–58.

30. Lee, B. and Richards, F. M. (1971). The interpretation of protein structures: estimation of static accessibility. *Journal of Molecular Biology*, **55**, 379–400.

31. Lesk, A. M. and Chothia, C. (1988). Elbow motion in the immunoglobulins involves a molecular ball-and-socket joint. *Nature*, **335**, 188–90.

32. Wyszecki, G. and Stiles, W. S. (1982). *Color Science: Concepts and Methods, Quantitative Data and Formulae* (2nd edn). Wiley, New York.

33. Astrua, M. (1973). *Manual and Color Reproduction for Printing and the Graphic Arts*. Fountain Press, Windsor.

34. Kueppers, H. (1982). *Color Atlas: A Practical Guide for Color Mixing* (trans. R. Marcinik). Barron's, Woodbury, New York.

35. Chothia, C. (1975). Structural invariants in protein folding. *Nature*, **254**, 304–8.

36. Chothia, C. (1984). Principles that determine the structures of proteins. *Annual Review of Biochemistry*, **53**, 537–72.

37. Iijima, H., Dunbar, J. B., Jr., and Marshall, G. (1987). Calibration of effective Van der Waals atomic contact radii for proteins and peptides. *Proteins: Structure, Functions and Genetics*, **2**, 330–9.

38. Katz, L. and Levinthal, C. (1966). Molecular model-building by computer. *Scientific American*, **214** (6), 42–52.

39. Preparata, F. P. and Shamos, M. I. (1985). *Computational Geometry: An Introduction*. Springer-Verlag, New York.

40. Rashin, A. A., Iofin, M., and Honig, B. (1986). Internal cavities and buried waters in globular proteins. *Biochemistry*, **24**, 3619–25.

41. Chothia, C., Levitt, M., and Richardson, D. (1977). Structure of proteins: Packing of α-helices and pleated sheets. *Proceedings of the National Academy of Sciences of the USA*, **74**, 4130–4.

42. Chothia, C., Levitt, M., and Richardson, D. (1981). Helix to helix packing in proteins. *Journal of Molecular Biology*, **145**, 215–50.

43. Shrake, A. and Rupley, J. A. (1972). Environment and exposure to solvent of protein atoms. Lysozyme and insulin. *Journal of Molecular Biology*, **79**, 351–71.

44. Sloane, N. J. A. and Conway, J. H. (1987). *Sphere Packing*. Springer-Verlag, New York.

45. Richardson, J. S. (1981). The anatomy and taxonomy of protein structure. *Advances in Protein Chemistry*, **34**, 167–339.

46. Lesk, A. M. and Hardman, K. D. (1982). Computer-generated schematic diagrams of protein structures. *Science*, **216**, 539–40.

47. Priestle, J. P. (1988). RIBBON: a stereo cartoon drawing program for proteins. *Journal of Applied Crystallography*, **21**, 572–6.

48. Richardson, J. S., Getzoff, E. D., and Richardson, D. (1978). The β-bulge: A common small unit of nonrepetitive protein structure. *Proceedings of the National Academy of Sciences of the USA*, **75**, 2574–8.

49. Levitt, M. and Greer, J. (1977). Automatic interpretation of secondary structure in globular proteins. *Journal of Molecular Biology*, **114**, 181–239.

50. Kabsch, W. and Sander, C. (1984). Dictionary of protein secondary structure: Pattern recognition of hydrogen-bonded and geometrical features. *Biopolymers*, **22**, 2577–637.

51. Lesk, A. M. (1988). Molecular evolution. In *Computational Molecular*

Biology: Sources and Methods for Sequence Analysis, pp. 205–15. Oxford University Press, Oxford.

52. Needleman, S. B. and Wunsch, C. D. (1970). A general method applicable to the search for similarities in the amino acid sequence of two proteins. *Journal of Molecular Biology*, **48**, 443–53.

53. Boswell, D. R. and Lesk, A. M. (1988). Sequence comparison and alignment: The measurement and interpretation of sequence similarity. In *Computational Molecular Biology: Sources and Methods for Sequence Analysis* (ed. A. M. Lesk), Ch. 14. Oxford University Press, Oxford.

54. Levitt, M. and Chothia, C. (1976). Structural patterns in globular proteins. *Nature*, **261**, 552–8.

55. Finkelstein, A. V. and Ptitsyn, O. B. (1978). Theory of self-organization of protein secondary structure—dependence of native globule structure on secondary structure of unfolded chain. *Doklady Akademiya Nauk SSSR*, **242**, 1226–8.

56. Klein, P., Kanehisa, M., and Delisi, C. (1984). Prediction of protein function from sequence properties: Discriminant analysis of a data base. *Biochimica et Biophysica Acta*, **787**, 221–6.

57. Klein, P., Kanehisa, M., and Delisi, C. (1985). The detection and classification of membrane-spanning proteins. *Biochimica et Biophysica Acta*, **815**, 468–76.

58. Klein, P. and Delisi, C. (1986). Prediction of protein structural class from the amino acid sequence. *Biopolymers*, **25**, 1659–72.

59. Bashford, D., Chothia, C., and Lesk, A. M. (1987). Determinants of a protein fold: Unique features of the globin amino acid sequences. *Journal of Molecular Biology*, **196**, 199–216.

60. Hodgeman, T. C. (1986). The elucidation of protein function from its amino acid sequence. *Computer Applications in the Biosciences*, **2**, 181–7.

61. Hodgeman, T. C. (1989). The elucidation of protein function by sequence motif analysis. *Computer Applications in the Biosciences*, **5**, 1–13.

62. Gould, S. J. (1983). *Hens' Teeth and Horses' Toes*, p. 46. Penguin Books, Harmondsworth.

63. Deisenhofer, J., Epp, O., Miki, K., Huber, R., and Michel, H. (1985). Structure of the protein subunits in the photosynthetic reaction centre of *Rhodopseudomonas viridis* at 3 Å resolution. *Nature*, **318**, 618–24.

64. Murzin, A. G. and Finkelstein, A. V. (1988). The general architecture of α-helical globules. *Journal of Molecular Biology*, **204**, 749–70.

65. Finkelstein, A. V. and Ptitsyn, O. B. (1987). Why do globular proteins fit the limited set of folding patterns? *Progress in Biophysics and Molecular Biology*, **50**, 171–90.

66. Brändén C.-I. (1980). Relation between structure and function of α/β-proteins. *Quarterly Reviews of Biophysics*, **13**, 317–38.

67. Tulinsky, A., Park, C. H., and Skzypczak-Jankun, E. (1988). Structure of prothombin fragment I at 2.8 Å resolution. *Journal of Molecular Biology*, **202**, 885–901.

68. Lesk, A. M., Brändén, C.-I., and Chothia, C. (1989). Structural principles of α–β barrel proteins: the packing of the interior of the sheet. *Proteins: Structure, Function and Genetics*, **5**, 139–48.

69. Goldman, A., Ollis, D. L., and Steitz, T. A. (1987). Crystal-structure of muconate lactonizing enzyme at 3 Å resolution. *Journal of Molecular Biology*, **194**, 143–53.

70. Stec, B. and Lebioda, L. (1990). Refined structure of yeast apo-enolase at 2.25 Å resolution. *Journal of Molecular Biology*, **211**, 235–48.

71. McLachlan, A. D. (1979). Gene duplications in the structural evolution of chymotrypsin. *Journal of Molecular Biology*, **128**, 49–79.

72. Chothia, C. and Lesk, A. M. (1987). The evolution of protein structures. *Cold Spring Harbor Symposia in Quantitative Biology*, LII, 399–405.

73. Rose, G. D., Young, W. B., and Gierasch, L. M. (1983). Interior turns in globular proteins. *Nature*, **304**, 654–7.

74. Rose, G. D., Gierasch, L. M., and Smith, John A. (1985). Turns in peptides and proteins. *Advances in Protein Chemistry*, **37**, 1–109.

75. Rose, G. D. (1978). Prediction of chain turns in globular proteins on a hydrophobic basis. *Nature*, **272**, 586–90.

76. Rose, G. D. (1988). Hydrophobicity profiles. In *Computational Molecular Biology: Sources and Methods for Sequence Analysis* (ed. A. M. Lesk), pp. 198–204. Oxford University Press, Oxford.

77. Crawford, I. P., Niermann, T., and Kirschner, K. (1987). Prediction of secondary structure by evolutionary comparison: Application to the α subunit of tryptophan synthase. *Proteins: Structure, Function and Genetics*, **2**, 118–29.

78. Venkatachalam, C. (1968). Stereochemical criteria for polypeptides and proteins. V. Conformation of a system of three linked peptide units. *Biopolymers*, **6**, 1425–36.

79. Sibanda, B. L. and Thornton, J. M. (1985). β-hairpin families in globular proteins. *Nature*, **316**, 170–4.

80. Efimov, A. V. (1986). Standard conformations of polypeptide chains in irregular regions of proteins. *Molecular Biology (USSR)*, **20**, 208–16.

81. Chothia, C. and Lesk, A. M. (1987). Canonical structures for the hypervariable regions of immunoglobulins. *Journal of Molecular Biology*, **196**, 901–18.

82. Wilmot, C. M. and Thornton, J. M. (1988). Analysis and prediction of the different types of β-turns in proteins. *Journal of Molecular Biology*, **203**, 221–32.

83. Sibanda, B. L., Blundell, T. L., and Thornton, J. M. (1989). Conformation of β-hairpins in protein structures. A systematic classification with applications to modeling by homology, electron density fitting and protein engineering. *Journal of Molecular Biology*, **206**, 759–77.

84. Leszczynski, J. F. and Rose, G. D. (1986). Loops in globular proteins: a novel category of secondary structure. *Science*, **234**, 849–55.

85. Jones, T. A. and Thirup, S. (1986). Use of known substructures in protein model building and crystallography. *EMBO Journal*, **5**, 819–22.

86. Tramontano, A., Chothia, C., and Lesk, A. M. (1989). Structural determinants of the conformations of medium-sized loops in proteins. *Proteins: Structure, Function and Genetics*, **6**, 382–94.

87. Chothia, C., Lesk, A. M., Tramontano, A., Levitt, M., Smith-Gill, S. J., Air, G., Sheriff, S., Padlan, E. A., Davies, D., Tulip, W. R., Colman, P. M., Spinelli, S., Alzari, P. M., and Poljak, R J. (1989). The conformations of immunoglobulin hypervariable regions. *Nature*, **342**, 877–83.

88. Bruccoleri, R. E., Haber, E., and Novotny, J. (1988). Structure of antibody hypervariable loops reproduced by a conformational search algorithm. *Nature*, **335**, 564–8.

89. Martin, A. C. R., Cheetham, J. C., and Rees, A. R. (1989). Modeling antibody hypervariable loops: a combined algorithm. *Proceedings of the National Academy of Sciences of the USA*, **86**, 9268–72.

90. Pflugrath, J. W. and Quiocho, F. A. (1988). The 2 Å resolution structure of the sulfate-binding protein involved in active transport in *Salmonella typhimurium*. *Journal of Molecular Biology*, **200**, 163–80.

91. Lesk, V. I. and Lesk, A. M. (1989). Schematic diagrams of nucleic acids and protein–nucleic acid complexes. *Journal of Applied Crystallography*, **22**, 569–71.

92. Schirmer, T. and Evans, P. R. (1990). Structural basis of the allosteric behaviour of phosphofructokinase. *Nature*, **343**, 140–5.

93. Golub, G. H. and Van Loan, C. F. (1983). *Matrix Computations*, pp. 425–6. Johns Hopkins University Press, Baltimore, Maryland.

94. Remington, S. J. and Matthews, B. W. (1978). A general method to assess similarity of protein structures, with applications to T4 bacteriophage lysozyme. *Proceedings of the National Academy of Sciences of the USA*, **75**, 2180–4.

95. Remington, S. J. and Matthews, B. W. (1980). A systematic approach to the comparison of protein structures. *Journal of Molecular Biology*, **140**, 77–99.

96. Levine, M., Stuart, D., and Williams, J. (1984). A method for systematic

comparison of the three-dimensional structures of proteins and some results. *Acta Crystallographica*, **A40**, 600–10.

97. Lesk, A. M. and Chothia, C. (1984). Mechanisms of domain closure in proteins. *Journal of Molecular Biology*, **174**, 175–91.

98. Perutz, M. F. (1989). *Mechanisms of cooperativity and allosteric regulation in proteins*. Cambridge University Press, Cambridge.

99. Phillips, S. E. V. (1980). Structure and refinement of oxymyoglobin at 1.6 Å resolution. *Journal of Molecular Biology*, **142**, 531–54.

100. Chothia, C., Lesk, A. M., Dodson, G. G., and Hodgkin, D. C. (1983). Transmission of conformational change in insulin. *Nature*, **302**, 500–5.

101. Remington, S. J., Wiegand, G., and Huber, R. (1982). Crystallographic refinement and atomic models of two different forms of citrate synthase at 2.7 and 1.7 Å resolution. *Journal of Molecular Biology*, **158**, 111–52.

102. Liddington, R., Derewenda, Z., Dodson, G., and Harris, D. (1988). Structure of the liganded T state of haemoglobin identifies the origin of cooperative oxygen binding. *Nature*, **331**, 725–8.

103. Monod, J., Wyman, J., and Changeux, J.-P. (1965). On the nature of allosteric transitions: A plausible model. *Journal of Molecular Biology*, **12**, 88–118.

104. Shaanan, B. (1983). Structure of human oxyhaemoglobin at 2.1 Å resolution. *Journal of Molecular Biology*, **171**, 31–51.

105. Fermi, G., Perutz, M. F., Shaanan, B., and Fourme, R. (1984). The crystal structure of human deoxyhaemoglobin at 1.74 Å resolution. *Journal of Molecular Biology*, **175**, 159–71.

106. Baldwin, J. and Chothia, C. (1979). Haemoglobin: The structural changes related to ligand binding and its allosteric mechanism. *Journal of Molecular Biology*, **129**, 175–220.

107. Gelin, B. R. and Karplus, M. (1977). Mechanism of tertiary structural change in hemoglobin. *Proceedings of the National Academy of Sciences of the USA*, **74**, 801–5.

108. Warshel, A. (1977). Energy-structure correlation in metalloporphyrins and the control of oxygen binding by hemoglobin. *Proceedings of the National Academy of Sciences of the USA*, **74**, 1789–93.

109. Chothia, C. and Lesk, A. M. (1985). Helix movements in proteins. *Trends in Biochemical Sciences*, **10**, 116–18 + cover + centrefold.

110. Lesk, A. M. and Chothia, C. (1986). The response of protein structures to amino acid sequence changes. *Philosophical Transactions of the Royal Society (London)*, **317**, 345–56.

111. Chothia, C. and Lesk, A. M. (1986). The relation between the divergence of sequence and structure in proteins. *EMBO Journal*, **5**, 823–6.

112. Krabbé, T. (1985). *Chess Curiosities*. Allen & Unwin, London.

113. Wells, J. A., Powers, D. B., Bott, R. R., Katz, B. A., Ultsch, M. H., Kossiakoff, A. A., Power, S. D., Adams, R. M., Heyneker, H. H., Cunningham, B. C., Miller, J. V., Graycar, T. P., and Estell, D. A. (1987). Protein engineering of subtilisin. In *Protein Engineering* (ed. D. L. Oxender and C. F. Fox), Alan R. Liss, New York.

114. Matsumura, M., Becktel, W. J., and Matthews, B. W. (1988). Hydrophobic stabilization in T4 lysozyme determined directly by multiple substitutions of Ile 3. *Nature*, **334**, 406–10.

115. Jones, T. A. (1985). Interactive computer graphics: FRODO. In *Methods in Enzymology*, Vol. 115 (ed. H. W. Wyckoff, C. H. W. Hirs, and S. N. Timasheff), pp. 157–71. Academic Press, New York.

116. Jones, T. A. (1987). Computer graphics in structure analysis. *Acta Crystallographica*, **A43** (Suppl.) ML18-1.

117. Unger, R., Harel, D., Wherland, S., and Sussman, J. L. (1989). A 3D building blocks approach to analyzing and predicting structure of proteins. *Proteins: Structure, Function and Genetics*, **5**, 355–73.

118. Dayringer, H. E., Tramontano, A., Sprang, S. R. and Fletterick, R. J. (1986). Interactive program for visualization and modelling of proteins, nucleic acids and small molecules. *Journal of Molecular Graphics*, **4**, 82–7.

119. Ferrin, T. E., Huang, C. C., Jarvis, L. E., and Langridge, R. (1988). The MIDAS database system. *Journal of Molecular Graphics*, **6**, 2–12.

120. Ferrin, T. E., Huang, C. C., Jarvis, L. E., and Langridge, R. (1988). The MIDAS display system. *Journal of Molecular Graphics*, **6**, 13–27.

121. Vriend, G. (1990). WHAT IF: a molecular modelling and drug design program. *Journal of Molecular Graphics*, **8**, 52–6.

122. Jones, T. A., Bergdoll, M., and Kjeldgaard, M. (1990). O: A macromolecule modeling environment. In *Crystallographic and Modeling Methods in Molecular Design* (eds. C. E. Bugg and S. E. Ealick), pp. 189–99. Springer-Verlag, New York.

123. Chothia, C. and Finkelstein, A. (1990). The classification and origins of protein folding patterns. *Annual Review of Biochemistry*, **59**, 1007–39.

124. Dirac, P. A. M. (1967). *The Principles of Quantum Mechanics*, p. 11. Oxford University Press, Oxford.

125. McRee, D. E., Tainer, J. A., Meyer, T. E., Van Beeumen, J., Cusanovitch, M. A., and Getzoff, E. D. (1989). Crystallographic structure of a photo-receptor protein at 2.4 Å resolution. *Proceedings of the National Academy of Sciences of the USA*, **86**, 6533–7.

Bibliography of protein structure determinations

The following publications correspond to the coordinate sets deposited in the Protein Data Bank (July 1990 release). Each entry is identified by a four-character code. The entries are sorted in order of the last three characters of the code. Citations containing authors and titles but no literature reference appear in the Data Bank as 'To be published'.

351C, 451C Y. Matsuura, T. Takano, and R. E. Dickerson (1982). Structure of cytochrome c_{551} from *P. aeruginosa* refined at 1.6 Å resolution and comparison of the two redox forms. *J. Mol. Biol.* **156**, 389.

155C R. Timkovich and R. E. Dickerson (1976). The structure of *Paracoccus denitrificans* cytochrome c_{550}. *J. Biol. Chem.* **251**, 4033.

156B F. Lederer, A. Glatigny, P. H. Bethge, H. D. Bellamy, and F. S. Mathews (1981). Improvement of the 2.5 Å resolution model of cytochrome b_{562} by redetermining the primary structure and using molecular graphics. *J. Mol. Biol.* **148**, 427.

1AAT V. N. Malaskevich, V. M. Kochkina, Iu. M. Torchinskii, and E. G. Arutiunian (1982). Oxoglutarate-induced conformational changes in cytosolic aspartate amino-transferase. *Dokl. Akad. Nauk SSSR* **267**, 1257.

2AAT D. L. Smith, S. C. Almo, M. D. Toney, and D. Range (1989). 2.8 Å resolution crystal structure of an active site mutant of aspartate aminotransferase from *Escherichia coli*. *Biochemistry* **28**, 8161.

1ABP G. L. Gilliland and F. A. Quiocho (1981). Structure of the L-arabinose-binding protein from *Escherichia coli* at 2.4 Å resolution. *J. Mol. Biol.* **146**, 341.

2ABX R. A. Love and R. M. Stroud (1986). The crystal structure of α-bungarotoxin at 2.5 Å resolution. Relation to solution structure and binding to acetylcholine receptor. *Protein Eng.* **1**, 37.

2ACT E. N. Baker and E. J. Dodson (1980). Crystallographic refinement of the structure of actinidin at 1.7 Å resolution by fast Fourier least-squares methods. *Acta Crystallogr.* **A36**, 559.

1ACX V. Z. Pletnev, A. P. Kuzin, and L. V. Malinina (1982). Actinoxanthin structure at the atomic level. *Bioorg. Khim.* **8**, 1637.

5ADH H. Eklund, J.-P. Samama, and T. A. Jones (1984). Crystallographic investigations of nicotinamide adenine dinucleotide binding to horse liver alcohol dehydrogenase. *Biochemistry* **23**, 5982.

6ADH H. Eklund, J.-P. Samama, L. Wallén, C.-I. Brändén, Å. Åkeson, and T. A. Jones (1981). Structure of triclinic ternary complex of horse liver alcohol dehydrogenase at 2.9 Å resolution. *J. Mol. Biol.* **146**, 561.

7ADH B. V. Plapp, H. Eklund, T. A. Jones, and C.-I. Brändén (1983). Three-dimensional structure of isonicotinimidylated liver alcohol dehydrogenase. *J. Biol. Chem.* **258**, 5537.

8ADH H. Eklund, J.-P. Samama, and T. A. Jones (1984). Crystallographic investigations of nicotinamide adenine dinucleotide binding to horse liver alcohol dehydrogenase. *Biochemistry* **23**, 5982.

3ADK D. Dreusicke, P. A. Karplus, and G. E. Schulz (1988). Refined structure of porcine cytosolic adenylate kinase at 2.1 Å resolution. *J. Mol. Biol.* **199**, 359.

1AGA S. Arnott, A. Fulmer, W. E. Scott, I. C. M. Dea, R. Moorhouse, and D. A. Rees (1974). The agarose double helix and its function in agarose gel structure. *J. Mol. Biol.* **90**, 269.

2AIT A. D. Kline, W. Braun, and K. Wuethrich (1988). Determination of the complete three-dimensional structure of the α-amylase inhibitor tendamistat in aqueous solution by nuclear magnetic resonance and distance geometry. *J. Mol. Biol.* **204**, 675.

1ALC K. R. Acharya, D. I. Stuart, N. P. C. Walker, M. Lewis, and D. C. Phillips (1989). Refined structure of baboon α-lactalbumin at 1.7 Å resolution. Comparison with c-type lysozyme. *J. Mol. Biol.* **208**, 99.

2ALP M. Fujinaga, L. T. J. Delbaere, G. D. Brayer, and M. N. G. James (1985). Refined structure of α-lytic protease at 1.7 Å resolution. Analysis of hydrogen bonding and solvent structure. *J. Mol. Biol.* **184**, 479.

1AMT R. O. Fox and F. M. Richards (1982). A voltage-gated ion channel and model inferred from the crystal structure of alamethicin at 1.5 Å resolution. *Nature* **300**, 325.

1ANA B. N. Conner, C. Yoon, J. L. Dickerson, and R. E. Dickerson (1984). Helix geometry and hydration in an A-DNA tetramer. [1]CCGG. *J. Mol. Biol.* **174**, 663.

2ANA M. McCall, T. Brown, and O. Kennard (1985). The crystal structure of d(G-G-G-G-C-C-C-C). A model for poly (dG)·poly(dC). *J. Mol. Biol.* **183**, 385.

3ANA H. Lauble, R. Frank, H. Bloecker, and U. Heinemann (1988). Three-dimensional structure of d(GGGATCCC) in the crystalline state. *Nucleic Acids Res.* **16**, 7799.

5ANA F. Takusagawa (1990). The crystal structure of d(GTACGTAC) at 2.25 Å resolution. Are the A-DNA's always unwound approximately 10 degrees at the C-G steps? *J. Biomol. Struct. Dyn.* **7**, 795.

4APE T. L. Blundell, J. A. Jenkins, B. T. Sewall, L. H. Pearl, J. B. Cooper, I. J. Tickle, B. Veerapandian, and S. P. Wood (1990). X-ray analyses of aspartic proteinases. The three-dimensional structure at 2.1 Å resolution of endothiapepsin. *J. Mol. Biol.* **211**, 919.

5API, 6API H. Loebermann, R. Tokuoka, J. Deisenhofer, and R. Huber (1984). Human α_1-proteinase inhibitor. Crystal structure analysis of two crystal modifications, molecular model and preliminary analysis of the implications for function. *J. Mol. Biol.* **177**, 531.

2APK I. T. Weber, T. A. Steitz, J. Bubis, and S. S. Taylor (1987). Predicted structures of cAMP binding domains of type I and II regulatory subunits of cAMP-dependent protein kinase. *Biochemistry* **26**, 343.

2APP M. N. G. James and A. R. Sielecki (1983). Structure and refinement of penicillopepsin at 1.8 Å resolution. *J. Mol. Biol.* **163**, 299.

Appendix 1

2APR K. Suguna, R. R. Bott, E. A. Padlan, E. Subramanian, S. Sheriff, G. H. Cohen, and D. R. Davies (1987). Structure and refinement at 1.8 Å resolution of the aspartic proteinase from *Rhizopus chinensis*. *J. Mol. Biol.* **196**, 877.

3APR K. Suguna, E. A. Padlan, C. W. Smith, W. D. Carlson, and D. R. Davies (1987). Binding of a reduced peptide inhibitor to the aspartic proteinase from *Rhizopus chinensis*. Implications for a mechanism of action. *Proc. Nat. Acad. Sci. USA* **84**, 7009.

2ATC R. B. Honzatko, J. L. Crawford, H. L. Monaco, J. E. Ladner, B. F. P. Edwards, D. R. Evans, S. G. Warren, D. C. Wiley, R. C. Ladner, and W. N. Lipscomb (1982). Crystal and molecular structures of native and CTP-liganded aspartate carbamoyltransferase from *Escherichia coli*. *J. Mol. Biol.* **160**, 219.

4ATC H.-M. Ke, R. B. Honzatko, and W. N. Lipscomb (1984). Structure of unligated aspartate carbamoyltransferase of *Escherichia coli* at 2.6-Å resolution. *Proc. Nat. Acad. Sci. USA* **81**, 4037.

7ATC K. H. Kim, Z. Pan, R. B. Honzatko, H. Ke, and W. N. Lipscomb (1987). Structural asymmetry in the CTP-liganded form of aspartate carbamoyltransferase from *Escherichia coli*. *J. Mol. Biol.* **196**, 853.

2AZA E. N. Baker (1988). Structure of azurin from *Alcaligenes denitrificans*. Refinement at 1.8 Å resolution and comparison of the two crystallographically independent molecules. *J. Mol. Biol.* **203**, 1071.

1AZU E. T. Adman and L. H. Jensen (1981). Structural features of azurin at 2.7 Å resolution. *Isr. J. Chem.* **21**, 8.

2B5C F. S. Matthews, P. Argos, and M. Levine (1972). The structure of cytochrome b_5 at 2.0 Å resolution. *Cold Spr. Harb. Symp. Quant. Biol.* **36**, 387.

3BCL D. E. Tronrud, M. F. Schmid, and B. W. Matthews (1986). Structure and x-ray amino acid sequence of a bacteriochlorophyll A protein from *Prosthecochloris aestuarii* refined at 1.9 Å resolution. *J. Mol. Biol.* **188**, 443.

1BD1 U. Heinemann and C. Alings (1989). Crystallographic study of one turn of G/C-rich B-DNA. *J. Mol. Biol.* **210**, 369.

1BDS, 2BDS P. C. Driscoll, A. M. Gronenborn, L. Beress, and G. M. Clore (1988). Determination of the three-dimensional solution structure of the antihypertensive and antiviral protein BDS-I from the sea anemone *Anemonia sulcata*. A study using nuclear magnetic resonance and hybrid distance geometry-dynamical simulated annealing. *Biochemistry* **28**, 2188.

1BLM O. Herzberg and J. Moult (1987). Bacterial resistance to β-lactam antibiotics. Crystal structure of β-lactamase from *Staphylococcus aureus* PC1 at 2.5 Å resolution. *Science* **236**, 694.

1BNA H. R. Drew, R. M. Wing, T. Takano, C. Broka, S. Tanaka, K. Itakura, and R. E. Dickerson (1981). Structure of a B-DNA dodecamer. Conformation and dynamics. *Proc. Nat. Acad. Sci. USA* **78**, 2179.

2BNA H. R. Drew, S. Samson, and R. E. Dickerson (1982). Structure of a B-DNA dodecamer at 16 kelvin. *Proc. Nat. Acad. Sci. USA* **79**, 4040.

3BNA, 4BNA A. V. Fratini, M. L. Kopka, H. R. Drew, and R. E. Dickerson (1982). Reversible bending and helix geometry in a B-DNA dodecamer. CGCGAATT$_{Br}$CGCG. *J. Biol. Chem.* **257**, 14686.

5BNA R. M. Wing, P. Pjura, H. R. Drew, and R. E. Dickerson (1984). The primary mode of binding of cisplatin to a B-DNA dodecamer. C-G-C-G-A-A-T-T-C-G-C-G. *EMBO J.* **3**, 1201.

6BNA M. L. Kopka, C. Yoon, D. Goodsell, P. Pjura, and R. E. Dickerson (1985). Binding of an antitumor drug to DNA. Netropsin and C-G-C-G-A-A-T-T-$_{Br}$C-G-C-G. *J. Mol. Biol.* **183**, 553.

7BNA S. R. Holbrook, R. E. Dickerson, and S.-H. Kim (1985). Anisotropic thermal-parameter refinement of the DNA dodecamer CGCGAATTCGCG by the segmented rigid-body method. *Acta Crystallogr.* **B41**, 255.

8BNA P. E. Pjura, K. Grzeskowiak, and R. E. Dickerson (1987). Binding of Hoechst 33258 to the minor groove of B-DNA. *J. Mol. Biol.* **197**, 257.

1BP2 B. W. Dijkstra, K. H. Kalk, W. G. J. Hol, and J. Drenth (1981). Structure of bovine pancreatic phospholipase A2 at 1.7 Å resolution. *J. Mol. Biol.* **147**, 97.

2BP2 B. W. Dijkstra, G. J. H. van Nes, K. H. Kalk, N. P. Brandenburg, W. G. J. Hol, and J. Drenth (1982). The structure of bovine pancreatic prophospholipase A$_2$ at 3.0 Å resolution. *Acta Crystallogr.* **B38**, 793.

3BP2 B. W. Dijkstra, K. H. Kalk, J. Drenth, G. H. de Haas, M. R. Egmond, and A. J. Slotboom (1984). Role of the N-terminus in the interaction of pancreatic phospholipase A$_2$ with aggregated substrates. Properties and crystal structure of transaminated phospholipase A$_2$. *Biochemistry* **23**, 2759.

2BPK I. T. Weber, T. A. Steitz, J. Bubis, and S. S. Taylor (1987). Predicted structures of cAMP binding domains of type I and II regulatory subunits of cAMP-dependent protein kinase. *Biochemistry* **26**, 343.

2C2C, 3C2C G. E. Bhatia (1981). Refinement of the crystal structure of oxidized *Rhodospirillum rubrum* cytochrome c2 Thesis, University of California, San Diego.

1C4S W. T. Winter, S. Arnott, D. H. Isaac, and E. D. T. Atkins (1978). Chondroitin-4-sulfate. The structure of a sulfated glycosaminoglycan. *J. Mol. Biol.* **125**, 1.

2C4S J. J. Cael, W. T. Winter, and S. Arnott (1978). Calcium chondroitin 4-sulfate. Molecular conformation and organization of polysaccharide chains in a proteoglycan. *J. Mol. Biol.* **125**, 21

1CA2 A. E. Eriksson, T. A. Jones, and A. Liljas (1988). Refined structure of human carbonic anhydrase II at 2.0 Å resolution. *Proteins: Struct., Funct., Genet.* **4**, 274.

2CA2, 3CA2 A. E. Eriksson, P. M. Kylsten, T. A. Jones, and A. Liljas (1988). Crystallographic studies of inhibitor binding sites in human carbonic anhydrase II. A pentacoordinated binding of the SCN$^-$ ion to the zinc at high pH. *Proteins: Struct., Funct., Genet.* **4**, 283.

2CAB K. K. Kannan, M. Ramanadham, and T. A. Jones (1984). Structure, refinement and function of carbonic anhydrase isozymes. Refinement of human carbonic anhydrase I. *Ann. N.Y. Acad. Sci.* **429**, 49.

1CAP R. Moorhouse, W. T. Winter, S. Arnott, and M. E. Bayer (1977). Conformation and molecular organization in fibers of the capsular polysaccharide from *Escherichia coli* M41 mutant. *J. Mol. Biol.* **109**, 373.

1CAR S. Arnott, W. E. Scott, D. A. Rees, and C. G. A. McNab (1974). ι-carrageenan. Molecular structure and packing of polysaccharide double helices in oriented fibres of divalent cation salts. *J. Mol. Biol.* **90**, 253.

4CAT B. K. Vainshtein, W. R. Melik-Adamyan, V. V. Barynin, A. A. Vagin, A. I. Grebenko, V. V. Borisov, K. S. Bartels, I. Fita, and M. G. Rossmann (1986). Three-dimensional structure of catalase from *Penicillium vitale* at 2.0 Å resolution. *J. Mol. Biol.* **188**, 49.

7CAT, 8CAT I. Fita and M. G. Rossmann (1985). The NADPH binding site on beef liver catalase. *Proc. Nat. Acad. Sci. USA* **82**, 1604.

1CBH, 2CBH P. J. Kraulis, G. M. Clore, M. Nilges, T. A. Jones, G. Pettersson, J. Knowles, and A. M. Gronenborn (1985). Determination of the three-dimensional structure of the C-terminal domain of cellobiohydrolase I from *Trichoderma reesei*. A study using nuclear magnetic resonance and hybrid distance geometry-dynamical simulated annealing. *Biochemistry* **28**, 7241.

1CBP J. M. Guss, E. A. Merritt, R. P. Phizackerly, B. Hedman, M. Murata, K. O. Hodgson, and H. C. Freeman (1988). Phase determination by multiple-wavelength x-ray diffraction. Crystal structure of a basic "blue" copper protein from cucumbers. *Science* **241**, 806.

1CC5 D. C. Carter, K. A. Melis, S. E. O'Donnell, B. K. Burgess, W. F. Furey Jr, B.-C. Wang, and C. D. Stout (1985). Crystal structure of *Azotobacter* cytochrome c_5 at 2.5 Å resolution. *J. Mol. Biol.* **184**, 279.

1CCR H. Ochi, Y. Hata, N. Tanaka, M. Kakudo, T. Sakurai, S. Aihara, and Y. Morita (1983). Structure of rice ferricytochrome c at 2.0 Å resolution. *J. Mol. Biol.* **166**, 407.

2CCY B. C. Finzel, P. C. Weber, K. D. Hardman, and F. R. Salemme (1985). Structure of ferricytochrome c' from *Rhodospirillum molischianum* at 1.67 Å resolution. *J. Mol. Biol.* **186**, 627.

2CDV Y. Higuchi, M. Kusunoki, Y. Matsuura, N. Yasuoka, and M. Kakudo (1984). Refined structure of cytochrome c_3 at 1.8 Å resolution. *J. Mol. Biol.* **172**, 109.

2CGA D. Wang, W. Bode, and R. Huber (1985). Bovine chymotrypsinogen a. X-ray crystal structure analysis and refinement of a new crystal form at 1.8 Å resolution. *J. Mol. Biol.* **185**, 595.

2CHA J. J. Birktoft and D. M. Blow (1972). The structure of crystalline α-chymotrypsin, V. The atomic structure of tosyl-α-chymotrypsin at 2 Å resolution. *J. Mol. Biol.* **68**, 187.

4CHA H. Tsukada and D. M. Blow (1985). Structure of α-chymotrypsin refined at 1.68 Å resolution. *J. Mol. Biol.* **184**, 703.

5CHA R. A. Blevins and A. Tulinsky (1985). The refinement and the structure of the dimer of α-chymotrypsin at 1.67-Å resolution. *J. Biol. Chem.* **260**, 4264.

6CHA A. Tulinsky and R. A. Blevins (1987). Structure of a tetrahedral transition state complex of α-chymotrypsin at 1.8-Å resolution. *J. Biol. Chem.* **262**, 7737.

1CHG S. T. Freer, J. Kraut, J. D. Robertus, H. T. Wright, and N. H. Xuong (1970). Chymotrypsinogen, 2.5 Å crystal structure, comparison with α-chymotrypsin, and implications for zymogen activation. *Biochemistry* **9**, 1997.

1CHO M. Fujinaga, A. R. Sielecki, R. J. Read, W. Ardelt, M. Laskowski, Jr., and M. N. G. James (1987). Crystal and molecular structures of the complex of α-chymotrypsin with its inhibitor turkey ovomucoid third domain at 1.8 Å resolution. *J. Mol. Biol.* **195**, 397.

2CI2 C. A. McPhalen and M. N. G. James (1987). Crystal and molecular structure of the serine proteinase inhibitor CI-2 from barley seeds. *Biochemistry* **26**, 261.

2CLN N. C. J. Strynadka and M. N. G. James (1988). Two trifluoperazine-binding sites on calmodulin predicted from comparative molecular modelling with troponin-c. *Proteins: Struct., Funct., Genet.* **3**, 1.

3CLN Y. S. Babu, C. E. Bugg, and W. J. Cook (1988). Structure of calmodulin refined at 2.2 Å resolution. *J. Mol. Biol.* **204**, 191.

1CMS G. L. Gilliland, E. L. Winborne, J. Nachman, and A. Wlodawer (1990). The three-dimensional structure of recombinant bovine chymosin at 2.3 Å resolution. *Proteins: Struct., Funct., Genet.* **8**, 82.

1CN1 M. Shoham, A. Yonath, J. L. Sussman, J. Moult, W. Traub, and A. J. Kalb (Gilboa) (1979). Crystal structure of demetallized concanavalin a. The metal-binding region. *J. Mol. Biol.* **131**, 137.

2CNA G. N. Reeke, Jr., J. W. Becker, and G. M. Edelman (1975). The covalent and three-dimensional structure of concanavalin a, IV. Atomic coordinates, hydrogen bonding, and quaternary structure. *J. Biol. Chem.* **250**, 1525.

3CNA K. D. Hardman and C. F. Ainsworth (1972). Structure of concanavalin a at 2.4 Å resolution. *Biochemistry* **11**, 4910.

1COH B. Luisi and N. Shibayama (1989). Structure of haemoglobin in the deoxy quaternary state with ligand bound at the α haems. *J. Mol. Biol.* **206**, 723.

3CPA D. W. Christianson and W. N. Lipscomb (1986). X-ray crystallographic investigation of substrate binding to carboxypeptidase a at subzero temperature. *Proc. Nat. Acad. Sci. USA* **83**, 7568.

4CPA D. C. Rees and W. N. Lipscomb (1982). Refined crystal structure of the potato inhibitor complex of carboxypeptidase a at 2.5 Å resolution. *J. Mol. Biol.* **160**, 475.

5CPA D. C. Rees, M. Lewis, and W. N. Lipscomb (1983). Refined crystal structure of carboxypeptidase a at 1.54 Å resolution. *J. Mol. Biol.* **168**, 367.

1CPB M. F. Schmid and J. R. Herriott (1976). Structure of carboxypeptidase b at 2.8 Å resolution. *J. Mol. Biol.* **103**, 175.

2CPP T. L. Poulos, B. C. Finzel, and A. J. Howard (1987). High-resolution crystal structure of cytochrome P450CAM. *J. Mol. Biol.* **195**, 687.

3CPP R. Raag and T. L. Poulos (1985). Crystal structure of the carbon monoxy-substrate-cytochrome $P450_{CAM}$ ternary complex. *Biochemistry* **28**, 7586.

1CPV, 2CPV, 3CPV P. C. Moews and R. H. Kretsinger (1975). Refinement of the structure of carp muscle calcium-binding parvalbumin by model building and difference Fourier analysis. *J. Mol. Biol.* **91**, 201.

1CRN M. M. Teeter (1984). Water structure of a hydrophobic protein at atomic resolution. Pentagon rings of water molecules in crystals of crambin. *Proc. Nat. Acad. Sci. USA* **81**, 6014.

1CRO Y. Takeda, J. G. Kim, C. G. Caday, E. Steers, Jr, D. H. Ohlendorf, W. F. Anderson, and B. W. Matthews (1986). Different interactions used by cro repressor in specific and nonspecific DNA binding. *J. Biol. Chem.* **261**, 8608.

2CRO A. Mondragon, C. Wolberger, and S. C. Harrison (1989). Structure of phage 434 cro protein at 2.35 Å resolution. *J. Mol. Biol.* **205**, 179.

1CSE W. Bode, E. Papamokos, and D. Musil (1987). The high-resolution x-ray crystal structure of the complex formed between subtilisin Carlsberg and eglin c, an elastase inhibitor from the leech *Hirudo medicinalis*. Structural analysis, subtilisin structure and interface geometry. *Eur. J. Biochem.* **166**, 673.

1CTF M. Leijonmarck and A. Liljas (1987). Structure of the C-terminal domain of the ribosomal protein L7/L12 from *Escherichia coli* at 1.7 Å. *J. Mol. Biol.* **195**, 555.

1CTS, 2CTS, 3CTS S. Remington, G. Wiegand, and R. Huber (1982). Crystallographic refinement and atomic models of two different forms of citrate synthase at 2.7 and 1.7 Å resolution. *J. Mol. Biol.* **158**, 111.

4CTS G. Wiegand, S. Remington, J. Deisenhofer, and R. Huber (1984). Crystal structure analysis and molecular model of a complex of citrate synthase with oxaloacetate and S-acetonyl-coenzyme A. *J. Mol. Biol.* **174**, 205.

1CTX M. D. Walkinshaw, W. Saenger, and A. Maelicke (1980). Three-dimensional structure of the "long" neurotoxin from cobra venom. *Proc. Nat. Acad. Sci. USA* **77**, 2400.

1CY3 M. Pierrot, R. Haser, M. Frey, F. Payan, and J.-P. Astier (1982). Crystal structure and electron transfer properties of cytochrome c_3. *J. Biol. Chem.* **257**, 14341.

1CYC N. Tanaka, T. Yamane, T. Tsukihara, T. Ashida, and M. Kakudo (1975). The crystal structure of bonito (katsuo) ferrocytochrome c at 2.3 Å resolution. II. Structure and function. *J. Biochem. (Tokyo)* **77**, 147.

2CYP B. C. Finzel, T. L. Poulos, and J. Kraut (1984). Crystal structure of yeast cytochrome c peroxidase refined at 1.7-Å resolution. *J. Biol. Chem.* **259**, 13027.

3CYT T. Takano and R. E. Dickerson (1980). Redox conformation changes in refined tuna cytochrome c. *Proc. Nat. Acad. Sci. USA* **77**, 6371.

5CYT T. Takano (1984). Refinement of myoglobin and cytochrome c. *Methods and applications in crystallographic computing*, p. 262 (S. R. Hall and T. Ashida, Eds). Oxford University Press, Oxford.

1D16 R. Chattopadhyaya, S. Ikuta, K. Grzeskowiak, and R. E. Dickerson.

1DCG R. V. Gessner, C. A. Frederick, G. J. Quigley, A. Rich, and A. H.-J. Wang (1989). The molecular structure of the left-handed Z-DNA double helix at 1.0-Å atomic resolution. Geometry, conformation, and ionic interactions of d(CGCGCG). *J. Biol. Chem.* **264**, 7921.

2DCG A. H.-J. Wang, G. J. Quigley, F. J. Kolpak, J. L. Crawford, J. H. van Boom, G. van der Marel, and A. Rich (1979). Molecular structure of a left-handed double helical DNA fragment at atomic resolution. *Nature* **282**, 680.

3DFR, 4DFR J. T. Bolin, D. J. Filman, D. A. Matthews, R. C. Hamlin, and J. Kraut (1982). Crystal structures of *Escherichia coli* and *Lactobacillus casei* dihydrofolate reductase refined at 1.7 Å resolution. I. General features and binding of methotrexate molecule in a one-to-one complex. *J. Biol. Chem.* **257**, 13650 (1982).

8DFR J. F. Davies II, D. A. Matthews, S. J. Oatley, B. T. Kaufman, N.-H. Xuong, and J. Kraut. Refined crystal structures of chicken liver dihydrofolate reductase. 3 Å apo-enzyme and 1.7 Å NADPH holo-enzyme complex.

2DHB W. Bolton and M. F. Perutz (1970). Three dimensional Fourier synthesis of horse deoxyhaemoglobin at 2.8 Å resolution. *Nature* **228**, 551.

1DN4, 1DN5 B. Chevrier, A. C. Dock, B. Hartmann, M. Leng, D. Moras, M. T. Thuong, and E. Westhof (1986). Solvation of the left-handed hexamer d(5BrC-G-5BrC-G-5BrC-G) in crystals grown at two temperatures. *J. Mol. Biol.* **188**, 707.

1DN6 M. McCall, T. Brown, W. N. Hunter, and O. Kennard (1986). The crystal structure of d(GGATGGGAG). An essential part of the binding site for transcription factor IIIA. *Nature* **322**, 661.

1DN7 M. McCall, T. Brown, and O. Kennard (1985). The crystal structure of d(G-G-G-G-C-C-C-C]. A model for poly (dG)·poly(dC). *J. Mol. Biol.* **183**, 385.

1DN8 R. G. Brennan, E. Westhof, and M. Sundaralingam (1986). Structure of a Z-DNA with two different backbone chain conformations. Stabilization of the decadeoxy-oligonucleotide d(CGTACGTACG) by $Co(NH_3)_6^{3+}$ binding to the guanine. *J. Biomol. Struct. Dyn.* **3**, 649.

1DN9 C. Yoon, G. G. Prive, D. S. Goodsell, and R. E. Dickerson (1988). Structure of an alternating-B-DNA helix and its relationship to A-tract DNA. *Proc. Nat. Acad. Sci. USA* **85**, 6332.

9DNA U. Heinemann, H. Lauble, R. Frank, and H. Bloecker. Crystal structure analysis of an A-DNA fragment at 1.8 Å resolution. d(GCCCGGGC)

3DNB G. G. Prive, U. Heinemann, S. Chandrasegaran, L.-S. Kan, M. L. Kopka, and R. E. Dickerson (1987). Helix geometry, hydration, and G. A. mismatch in a B-DNA decamer. *Science* **238**, 498.

4DNB C. A. Frederick, G. J. Quigley, G. A. van der Marel, J. H. van Boom, A. H.-J. Wang, and A. Rich (1988). Methylation of the ecoRI recognition site does not alter DNA conformation. The crystal structure of $d(GCGCAm_6ATTCGCG)$ at 2.0-Å resolution. *J. Biol. Chem.* **263**, 17872.

2DND M. Coll, C. A. Frederick, A. H.-J. Wang, and A. Rich (1987). A bifurcated hydrogen-bonded conformation in the d(A·T) base pairs of the DNA dodecamer d(CGCAAATTTGCG) and its complex with distamycin. *Proc. Nat. Acad. Sci. USA* **84**, 8385.

1DNE M. Coll, J. Aymami, G. A. Van der Marel, J. H. Van Boom, A. Rich, and A. H.-J. Wang (1989). Molecular structure of the netropsin-d(CGCGATATCGCG) complex. DNA conformation in an alternating AT segment. *Biochemistry* **28**, 310.

1DNH M.-K. Teng, N. Usman, C. A. Frederick, and A. H.-J. Wang (1988). The molecular structure of the complex of Hoechst 33258 and the DNA dodecamer d(CGCGAATTCGCG). *Nucleic Acids Res.* **16**, 2671.

1DNN E. N. Trifonov and J. L. Sussman (1989). Smooth bending of DNA in chromatin. *Molecular Mechanisms of Biological Recognition.* (Ed. M. Balaban). Elsevier/North-Holland Biomedical Press, New York, 1989. p. 227.

1DNS S. Jain, G. Zon, and M. Sundaralingam (1989). Base only binding of spermine in the deep groove of the A-DNA octamer d(GTGTACAC). *Biochemistry* **28**, 2360.

1DPI D. L. Ollis, P. Brick, R. Hamlin, N. G. Xuong, and T. A. Steitz (1985). Structure of larger fragment of *Escherichia coli* DNA polymerase I complexed with dTMP. *Nature* **313**, 762.

3EBX J. L. Smith, P. W. R. Corfield, W. A. Hendrickson, and B. W. Low. Refinement at 1.4 Å resolution of a model of erabutoxin b. Treatment of ordered solvent and discrete disorder.

5EBX P. W. R. Corfield, T.-J. Lee, and B. W. Low (1989). The crystal structure of erabutoxin a at 2.00 Å resolution. *J. Biol. Chem.* **264**, 9239.

1ECA, 1ECD, 1ECN, 1ECO W. Steigemann and E. Weber (1979). Structure of erythrocruorin in different ligand states refined at 1.4 Å resolution. *J. Mol. Biol.* **127**, 309.

1EFM F. Jurnak (1985). Structure of the GDP domain of EFTU and location of the amino acids homologous to *ras* oncogene proteins. *Science* **230**, 32.

2ENL L. Lebioda, B. Stec, and J. M. Brewer (1989). The structure of yeast enolase at 2.25-Å resolution. An 8-fold $\beta + \alpha$-barrel with a novel $\beta\beta\alpha\alpha(\beta\alpha)_6$ topology. *J. Biol. Chem.* **264**, 3685.

1EST L. Sawyer, D. M. Shotton, J. W. Campbell, P. L. Wendell, H. Muirhead, H. C. Watson, R. Diamond, and R. C. Ladner (1978). The atomic structure of crystalline porcine pancreatic elastase at 2.5 Å resolution. Comparisons with the structure of α-chymotrypsin. *J. Mol. Biol.* **118**, 137.

2EST D. L. Hughes, L. C. Sieker, J. Bieth, and J.-L. Dimicoli (1982). Crystallographic study of the binding of a trifluoroacetyl dipeptide anilide inhibitor with elastase. *J. Mol. Biol.* **162**, 645.

3EST E. Meyer, G. Cole, R. Radahakrishnan, and O. Epp (1988). Structure of native porcine pancreatic elastase at 1.65 Å resolution. *Acta Crystallogr.* **B44**, 26.

1ETU T. F. M. la Cour, J. Nyborg, S. Thirup, and B. F. C. Clark (1985). Structural details of the binding of guanosine diphosphate to elongation factor Tu from *E. coli* as studied by x-ray crystallography. *EMBO J.* **4**, 2385.

1F19 M.-B. Lascombe, P. M. Alzari, G. Boulot, P. Saludjian, P. Tougard, C. Berek, S. Haba, E. M. Rosen, A. Nisonoff, and R. J. Poljak (1989). Three-dimensional structure of Fab R19.9, a monoclonal murine antibody specific for the *p*-azobenzenearsonate group. *Proc. Nat. Acad. Sci. USA* **86**, 607.

3FAB F. A. Saul, L. M. Amzel, and R. J. Poljak (1978). Preliminary refinement and structural analysis of the Fab fragment from human immunoglobulin NEW at 2.0 Å resolution. *J. Biol. Chem.* **253**, 585.

2FB4 M. Marquart, J. Deisenhofer, R. Huber, and W. Palm (1980). Crystallographic refinement and atomic models of the intact immunoglobulin molecule KOL and its antigen-binding fragment at 3.0 Å and 1.9 Å resolution. *J. Mol. Biol.* **141**, 369.

1FBJ S. W. Suh, T. N. Bhat, M. A. Navia, G. H. Cohen, D. N. Rao, S. Rudikoff, and D. R. Davies (1986). The galactan-binding immunoglobulin Fab J539. An X-ray diffraction study at 2.6-Å resolution. *Proteins: Struct., Funct., Genet.* **1**, 74.

1FC1, 1FC2 J. Deisenhofer (1981). Crystallographic refinement and atomic models of a human Fc fragment and its complex with fragment b of protein a from *Staphylococcus aureus* at 2.9- and 2.8-Å resolution. *Biochemistry* **20**, 2361.

4FD1 C. D. Stout (1989). Refinement of 7 Fe ferredoxin at 1.9 Å resolution. *J. Mol. Biol.* **205**, 545.

1FD2 A. Martin, B. K. Burgess, C. D. Stout, V. Cash, D. R. Dean, G. M. Jensen, and P. J. Stephens (1990). Site-directed mutagenesis of *Azotobacter vinelandii* ferredoxin I. (Fe-S) cluster driven protein rearrangement. *Proc. Nat. Acad. Sci. USA* **87**, 598.

1FDH J. A. Frier and M. F. Perutz (1977). Structure of human foetal deoxyhaemoglobin. *J. Mol. Biol.* **112**, 97.

1FDX E. T. Adman, L. C. Sieker, and L. H. Jensen (1976). Structure of *Peptococcus aerogenes* ferredoxin, refinement at 2 Å resolution. *J. Biol. Chem.* **251**, 3801.

1FVB, 2FVB, 1FVW, 2FVW E. A. Padlan and E. A. Kabat (1988). Model-building study of the combining sites of two antibodies to $\alpha(1\rightarrow6)$dextran. *Proc. Nat. Acad. Sci. USA* **85**, 6885.

1FX1 K. D. Watenpaugh, L. C. Sieker, and L. H. Jensen (1976). A crystallographic structural study of the oxidation states of *Desulfovibrio vulgaris* flavodoxin. *Flavins and flavoproteins*, p. 405 (Ed. T. P. Singer). Elsevier Scientific Publ. Co., Amsterdam.

1FXB K. Fukuyama, Y. Nagahara, T. Tsukihara, Y. Katsube, T. Hase, and H. Matsubara (1988). Tertiary structure of *Bacillus thermoproteolyticus* (4Fe-4S) ferredoxin. Evolutionary implications for bacterial ferredoxins. *J. Mol. Biol.* **199**, 183.

3FXC T. Tsukihara, K. Fukuyama, M. Nakamura, Y. Katsube, N. Tanaka, M. Kakudo, K. Wada, T. Hase, and H. Matsubara (1981). X-ray analysis of a (2Fe-2S) ferredoxin from *Spirulina platensis*. Main chain fold and location of side chains at 2.5 Å resolution. *J. Biochem. (Tokyo)* **90**, 1763.

3FXN, 4FXN W. W. Smith, R. M. Burnett, G. D. Darling, and M. L. Ludwig (1977). Structure of the semiquinone form of flavodoxin from *Clostridium MP*. Extension of 1.8 Å resolution and some comparisons with the oxidized state. *J. Mol. Biol.* **117**, 195.

2GAP I. T. Weber and T. A. Steitz (1984). Model of specific complex between catabolite gene activator protein and B-DNA suggested by electrostatic complementarity. *Proc. Nat. Acad. Sci. USA* **81**, 3973.

3GAP I. T. Weber and T. A. Steitz (1987). Structure of a complex of catabolite gene activator protein and cyclic AMP refined at 2.5 Å resolution. *J. Mol. Biol.* **198**, 311.

1GBP S. L. Mowbray and G. A. Petsko (1983). The x-ray structure of the periplasmic galactose binding protein from *Salmonella typhimurium* at 3.0 Å resolution. *J. Biol. Chem.* **258**, 7991.

2GCH G. H. Cohen, E. W. Silverton, and D. R. Davies (1981). Refined crystal structure of γ-chymotrypsin at 1.9 Å resolution. *J. Mol. Biol.* **148**, 449.

1GCN K. Sasaki, S. Dockerill, D. A. Adamiak, I. J. Tickle, and T. Blundell (1975). X-ray analysis of glucagon and its relationship to receptor binding. *Nature* **257**, 751.

1GCR L. Summers, G. Wistow, M. Narebor, D. Moss, P. Lindley, C. Slingsby, T. Blundell, H. Bartunik, and K. Bartels (1984). X-ray studies of the lens specific proteins. The crystallins. *Pept. Protein Rev.* **3**, 147.

1GD1 T. Skarzynski, P. C. E. Moody, and A. J. Wonacott (1987). Structure of holo-glyceraldehyde-3-phosphate dehydrogenase from *Bacillus stearothermophilus* at 1.8 Å resolution. *J. Mol. Biol.* **193**, 171.

2GD1 T. Skarzynski and A. J. Wonacott (1988). Coenzyme-induced conformational changes in glyceraldehyde-3-phosphate dehydrogenase from *Bacillus stearothermophilus*. *J. Mol. Biol.* **203**, 1097.

1GF1, 1GF2 T. L. Blundell, S. Bedarkar, and R. E. Humbel (1983). Tertiary structures, receptor binding, and antigenicity of insulinlike growth factors. *Fed. Proc.* **42**, 2592.

2GLS M. M. Yamashita, R. J. Almassy, C. A. Janson, D. Cascio, and D. Eisenberg (1989). Refined atomic model of glutamine synthetase at 3.5 Å resolution. *J. Biol. Chem.* **264**, 17681.

2GN5 G. D. Brayer and A. McPherson (1983). Refined structure of the gene 5 DNA binding protein from bacteriophage *fd*. *J. Mol. Biol.* **169**, 565.

1GOX Y. Lindqvist (1989). Refined structure of spinach glycolate oxidase at 2 Å resolution. *J. Mol. Biol.* **209**, 151.

1GP1 O. Epp, R. Ladenstein, and A. Wendel (1983). The refined structure of the selenoenzyme glutathione peroxidase at 0.2-nm resolution. *Eur. J. Biochem.* **133**, 51.

1GPD D. Moras, K. W. Olsen, M. N. Sabesan, M. Buehner, G. C. Ford, and M. G. Rossmann (1975). Studies of asymmetry in the three-dimensional structure of lobster D-glyceraldehyde-3-phosphate dehydrogenase. *J. Biol. Chem.* **250**, 9137.

3GPD W. D. Mercer, S. I. Winn, and H. C. Watson (1976). Twinning in crystals of human skeletal muscle D-glyceraldehyde-3-phosphate dehydrogenase. *J. Mol. Biol.* **104**, 277.

4GPD M. R. N. Murthy, R. M. Garavito, J. E. Johnson, and M. G. Rossmann (1980). The structure of lobster apo-D-glyceraldehyde-3-phosphate dehydrogenase at 3.0 Å resolution. *J. Mol. Biol.* **138**, 859.

3GRS P. A. Karplus and G. E. Schulz (1987). Refined structure of glutathione reductase at 1.54 Å resolution. *J. Mol. Biol.* **195**, 701.

1HBS E. A. Padlan and W. E. Love (1985). Refined crystal structure of deoxyhemoglobins. I. Restrained least-squares refinement at 3.0 Å resolution. *J. Biol. Chem.* **260**, 8272.

1HCO, 2HCO J. M. Baldwin (1980). The structure of human carbonmonoxy haemoglobin at 2.7 Å resolution. *J. Mol. Biol.* **136**, 103.

1HDS R. L. Girling, T. E. Houston, W. C. Schmidt, Jr, and E. L. Amma (1980). Macromolecular structure refinement by restrained least-squares and interactive graphics as applied to sickling deer type III hemoglobin. *Acta Crystallogr.* **A36**, 43.

2HFL S. Sheriff, E. W. Silverton, E. A. Padlan, G. H. Cohen, S. J. Smith-Gill, B. C. Finzel, and D. R. Davies (1987). Three-dimensional structure of an antibody-antigen complex. *Proc. Nat. Acad. Sci. USA* **84**, 8075.

1HFM S. J. Smith-Gill, C. Mainhart, T. B. Lavoie, R. J. Feldmann, W. Drohan, and B. R. Brooks (1987). A three-dimensional model of an anti-lysozyme antibody. *J. Mol. Biol.* **194**, 713.

2HFM S. J. Smith-Gill, C. Mainhart, B. R. Brooks. A model of an antibody-protein complex.

3HFM E. A. Padlan, E. W. Silverton, S. Sheriff, G. H. Cohen, S. J. Smith-Gill, and D. R. Davies (1989). Structure of an antibody-antigen complex. Crystal structure of the HyHEL-10 Fab-lysozyme complex. *Proc. Nat. Acad. Sci. USA* **86**, 5938.

2HHB, 3HHB, 4HHB G. Fermi, M. F. Perutz, B. Shaanan, and R. Fourme (1984). The crystal structure of human deoxyhaemoglobin at 1.74 Å resolution. *J. Mol. Biol.* **175**, 159.

1HHO B. Shaanan (1983). Structure of human oxyhaemoglobin at 2.1 Å resolution. *J. Mol. Biol.* **171**, 31.

1HIP C. W. Carter, Jr., J. Kraut, S. T. Freer, N.-H. Xuong, R. A. Alden, and R. G. Bartsch (1974). Two-Å crystal structure of oxidized chromatium high potential iron protein. *J. Biol. Chem.* **249**, 4212.

2HIR, 4HIR, 5HIR, 6HIR P. J. M. Folkers, G. M. Clore, P. C. Driscoll, J. Dodt, S.

Koehler, and A. M. Gronenborn (1989). Solution structure of recombinant hirudin and the lys-47→glu mutant. A nuclear magnetic resonance and hybrid distance geometry-dynamical simulated annealing study. *Biochemistry* **28**, 2601.

1HKG T. A. Steitz, M. Shoham, and W. S. Bennett, Jr (1981). Structural dynamics of yeast hexokinase during catalysis. *Philos. Trans. Roy. Soc. London*, **B29**, 43.

1HLA P. J. Bjorkman, M. A. Spaer, B. Samraoui, W. S. Bennett, J. L. Strominger, and D. C. Wiley (1987). Structure of the human class I histocompatibility antigen, HLA-A2. *Nature* **329**, 506.

2HLA T. P. J. Garrett, M. A. Saper, P. J. Bjorkman, J. L. Strominger, and D. C. Wiley (1989). Specificity pockets for the side chains of peptide antigens in HLA-Aw68. *Nature* **342**, 692.

3HLA M. A. Saper, P. J. Bjorkman, and D. C. Wiley (1991). Refined structure of the human histocompatibility antigen HLA-A2 at 2.6 Å resolution. *J. Mol. Biol.* **219**, 277.

1HMG I. A. Wilson, J. Skehel, and D. C. Wiley (1981). Structure of the haemagglutinin membrane glycoprotein of influenza virus at 3 Å resolution. *Nature* **289**, 366.

1HMQ, 1HMZ R. E. Stenkamp, L. C. Sieker, and L. H. Jensen (1983). Adjustment of restraints in the refinement of methemerythrin and azidomethemerythrin at 2.0 Å resolution. *Acta Crystallogr.* **B39**, 697.

1HNE M. A. Navia, B. M. McKeever, J. P. Springer, T.-Y. Lin, H. R. Williams, E. M. Fluder, C. P. Dorn, and K. Hoogsteen (1989). Structure of human neutrophil elastase in complex with a peptide chloromethyl ketone inhibitor at 1.84-Å resolution. *Proc. Nat. Acad. Sci. USA* **86**, 7.

1HOE J. W. Pflugrath, G. Wiegand, R. Huber, and L. Vertesy (1986). Crystal structure determination, refinement and the molecular model of the α-amylase inhibitor Hoe-467a. *J. Mol. Biol.* **189**, 383.

1HR3 J. L. Smith, W. A. Hendrickson, and A. W. Addison (1983). Structure of trimeric haemerythrin. *Nature* **303**, 86.

1HRB W. A. Hendrickson and K. B. Ward (1975). Atomic models for the polypeptide backbones of myohemerythrin and hemerythrin. *Biochem. Biophys. Res. Comm.* **66**, 1349.

1HVP I. T. Weber, M. Miller, M. Jaskolski, J. Leis, A. M. Skalka, and A. Wlodawer (1989). Molecular modeling of the HIV-1 protease and its substrate binding site. *Science* **243**, 928.

2HVP M. A. Navia, P. M. D. Fitzgerald, B. M. McKeever, C.-T. Leu, J. C. Heimbach, W. K. Herber, I. S. Sigal, P. L. Darke, and J. P. Springer (1989). Three-dimensional structure of aspartyl protease from human immunodeficiency virus HIV-1. *Nature* **337**, 615.

3HVP A. Wlodawer, M. Miller, M. Jaskolski, B. K. Sathyanarayana, E. Baldwin, I. T. Weber, L. M. Selk, L. Clawson, J. Schneider, and S. B. H. Kent (1989). Conserved folding in retroviral protease. Crystal structure of a synthetic HIV-1 protease. *Science* **245**, 616.

4HVP M. Miller, J. Schneider, B. K. Sathyanarayana, M. V. Toth, G. R. Marshall, L. Clawson, L. Selk, S. B. H. Kent, and A. Wlodawer (1989). Structure of complex of synthetic HIV-1 protease with a substrate-based inhibitor at 2.3 Å resolution. *Science* **246**, 1149.

1HYA W. T. Winter, P. J. C. Smith, and S. Arnott (1975). Hyaluronic acid, structure of a fully extended 3-fold helical sodium salt and comparison with the less extended 4-fold helical forms. *J. Mol. Biol.* **99**, 219.

2HYA, 3HYA J. M. Guss, D. W. L. Hukins, P. J. C. Smith, W. T. Winter, S. Arnott, R. Moorhouse, and D. A. Rees (1975). Hyaluronic acid, molecular conformations and interactions in two sodium salts. *J. Mol. Biol.* **95**, 359.

4HYA W. T. Winter and S. Arnott (1977). Hyaluronic acid, the role of divalent cations in conformation and packing. *J. Mol. Biol.* **117**, 761.

1I1B B. C. Finzel, L. L. Clancy, D. R. Holland, S. W. Muchmore, K. D. Watenpaugh, and H. M. Einspahr (1989). The crystal structure of recombinant human interleukin-1β at 2.0 Å resolution. *J. Mol. Biol.* **209**, 779.

2I1B J. P. Priestle, H.-P. Schaer, and M. G. Gruetter (1989). Crystallographic refinement of interleukin-1β at 2.0 Å resolution. *Proc. Nat. Acad. Sci. USA* **86**, 9667.

4I1B G. L. Gilliland, E. L. Winborne, Y. Masui, and Y. Hirai (1987). A preliminary crystallographic study of recombinant human interleukin-1β. *J. Biol. Chem.* **262**, 12323.

3ICB D. M. E. Szebenyi and K. Moffat (1986). The refined structure of vitamin D-dependent calcium-binding protein from bovine intestine. Molecular details, ion binding, and implications for the structure of other calcium-binding proteins. *J. Biol. Chem.* **261**, 8761.

2IG2 M. Marquart, J. Deisenhofer, R. Huber, and W. Palm (1980). Crystallographic refinement and atomic models of the intact immunoglobulin molecule KOL and its antigen-binding fragment at 3.0 Å and 1.9 Å resolution. *J. Mol. Biol.* **141**, 369.

1IGE E. A. Padlan and D. R. Davies (1986). A model of the Fc of immunoglobulin E. *Molecular Immunology* **23**, 1063.

1INS J. Bordas, G. G. Dodson, H. Grewe, M. H. J. Koch, B. Krebs, and J. Randall (1983). A comparative assessment of the zinc-protein coordination in 2Zn-insulin as determined by x-ray absorption fine structure (EXAFS) and x-ray crystallography. *Proc. Roy. Soc. London* **B219**, 21.

2INS G. D. Smith, W. L. Duax, E. J. Dodson, G. G. Dodson, R. A. G. de Graaf, and C. D. Reynolds (1982). The structure of des-phe B1 bovine insulin. *Acta Crystallogr.* **B38**, 3028.

3INS A. Wlodawer, H. Savage, and G. Dodson (1989). Structure of insulin. Results of joint neutron and x-ray refinement. *Acta Crystallogr.* **B45**, 99.

4INS E. N. Baker, T. L. Blundell, J. F. Cutfield, S. M. Cutfield, E. J. Dodson, G. G. Dodson, D. M. Crowfoot Hodgkin, R. E. Hubbard, N. W. Isaacs, C. D. Reynolds, K. Sakabe, N. Sakabe, and N. M. Vijayan (1988). The structure of 2Zn pig insulin crystals at 1.5 Å resolution. *Philos. Trans. R. Soc. London* **B319**, 369.

2KAI Z. Chen and W. Bode (1983). Refined 2.5 Å x-ray crystal structure of the complex formed by porcine kallikrein a and the bovine pancreatic trypsin inhibitor. Crystallization, Patterson search, structure determination, refinement, structure and comparison with its components and with the bovine trypsin-pancreatic trypsin inhibitor complex. *J. Mol. Biol.* **164**, 283.

1KES S. Arnott, J. M. Guss, D. W. L. Hukins, I. C. M. Dea, and D. A. Rees (1974). Conformation of keratan sulphate. *J. Mol. Biol.* **88**, 175.

1KGA I. M. Mavridis, M. H. Hatada, A. Tulinsky, and L. Lebioda (1982). Structure of 2-

keto-3-deoxy-6-phosphogluconate aldolase at 2.8 Å resolution. *J. Mol. Biol.* **162**, 419.

1L01 M. G. Gruetter, T. M. Gray, L. H. Weaver, T. Alber, K. Wilson, and B. W. Matthews (1987). Structural studies of mutants of the lysozyme of bacteriophage T4. The temperature-sensitive mutant protein Thr157→Ile. *J. Mol. Biol.* **197**, 315.

1L02, 1L03, 1L04, 1L05, 1L06, 1L07, 1L08, 1L09 T. Alber, S. Dao-Pin, K. Wilson, J. A. Wozniak, S. P. Cook, and B. W. Matthews (1987). Contributions of hydrogen bonds of Thr 157 to the thermodynamic stability of phage T4 lysozyme. *Nature* **330**, 41.

1L10 M. G. Gruetter, T. M. Gray, L. H. Weaver, T. Alber, K. Wilson, and B. W. Matthews (1987). Structural studies of mutants of the lysozyme of bacteriophage T4. The temperature-sensitive mutant protein Thr157→Ile. *J. Mol. Biol.* **197**, 315.

1L11, 1L12, 1L13, 1L14, 1L15 T. Alber, S. Dao-Pin, K. Wilson, J. A. Wozniak, S. P. Cook, and B. W. Matthews (1987). Contributions of hydrogen bonds of Thr 157 to the thermodynamic stability of phage T4 lysozyme. *Nature* **330**, 41.

1L16 T. M. Gray and B. W. Matthews (1987). Structural analysis of the temperature-sensitive mutant of bacteriophage T4 lysozyme, glycine 156→aspartic acid. *J. Biol. Chem.* **262**, 16858.

1L17, 1L18 M. Matsumura, W. J. Becktel, and B. W. Matthews (1988). Hydrophobic stabilization in T4 lysozyme determined directly by multiple substitutions of Ile 3. *Nature* **334**, 406.

1L19. 1L20 H. Nicholson, W. J. Becktel, and B. W. Matthews (1988). Enhanced protein thermostability from designed mutations that interact with α-helix dipoles. *Nature* **336**, 651.

1L21, 1L22 H. Nicholson, E. Soderlind, D. E. Tronrud, and B. W. Matthews (1989). Contributions of left-handed helical residues to the structure and stability of bacteriophage T4 lysozyme. *J. Mol. Biol.* **210**, 181.

1L23, 1L24 B. W. Matthews, H. Nicholson, and W. J. Becktel (1987). Enhanced protein thermostability from site-directed mutations that decrease the entropy of unfolding. *Proc. Nat. Acad. Sci. USA* **84**, 6663.

1L25, 1L26, 1L27, 1L28, 1L29, 1L30, 1L31, 1L32 T. Alber, J. A. Bell, S. Dao-Pin, H. Nicholson, J. A. Wozniak, S. Cook, and B. W. Matthews (1988). Replacements of Pro_{86} in phage T4 lysozyme extend an α-helix but do not alter protein stability. *Science* **239**, 631.

1L33 H. Nicholson, E. Soderlind, D. E. Tronrud, and B. W. Matthews (1989). Contributions of left-handed helical residues to the structure and stability of bacteriophage T4 lysozyme. *J. Mol. Biol.* **210**, 181.

1L34 L. H. Weaver, T. M. Gray, M. G. Gruetter, D. E. Anderson, J. A. Wozniak, F. W. Dahlquist, and B. W. Matthews (1989). High-resolution structure of the temperature-sensitive mutant of phage lysozyme, arg 96→his. *Biochemistry* **28**, 3793.

1L35 P. E. Pjura, M. Matsumura, J. A. Wozniak, and B. W. Matthews (1990). Structure of a thermostable disulfide-bridge mutant of phage T4 lysozyme shows that an engineered cross-link in a flexible region does not increase the rigidity of the folded protein. *Biochemistry* **29**, 2592.

1LDB, 2LDB K. Piontek, P. Chakrabarti, H.-P. Schaer, M. G. Rossmann, and H. Zuber (1990). Structure determination and refinement of *Bacillus stearothermophilus* lactate dehydrogenase. *Proteins: Struct., Funct., Genet.* **7**, 74.

3LDH J. L. White, M. L. Hackert, M. Buehner, M. J. Adams, G. C. Ford, P. J. Lentz, Jr., I. E. Smiley, S. J. Steindel, and M. G. Rossmann (1976). A comparison of the structures of apo dogfish M_4 lactate dehydrogenase and its ternary complexes. *J. Mol. Biol.* **102**, 759.

5LDH U. M. Grau, W. E. Trommer, and M. G. Rossmann (1981). Structure of the active ternary complex of pig heart lactate dehydrogenase with S-lac-NAD at 2.7 Å resolution. *J. Mol. Biol.* **151**, 289.

6LDH, 8LDH, 1LDM C. Abad-Zapatero, J. P. Griffith, J. L. Sussman, and M. G. Rossmann (1987). Refined crystal structure of dogfish M_4 apo-lactate dehydrogenase. *J. Mol. Biol.* **198**, 445.

2LDX H. H. Hogrefe, J. P. Griffith, M. G. Rossmann, and E. Goldberg (1987). Characterization of the antigenic sites on the refined 3-Å resolution structure of mouse testicular lactate dehydrogenase c_4. *J. Biol. Chem.* **262**, 13155.

1LH1, 2LH1, 1LH2, 2LH2, 1LH3, 2LH3, 1LH4, 2LH4, 1LH5, 2LH5, 1LH6, 2LH6, 1LH7, 2LH7 E. G. Arutyunyan, I. P. Kuranova, B. K. Vainshtein, and W. Steigemann (1980). X-ray structural investigation of leghemoglobin. VI. Structure of acetate-ferrileghemoglobin at a resolution of 2.0Å (Russian). *Kristallografiya* **25**, 80.

2LHB R. B. Honzatko, W. A. Hendrickson, and W. E. Love (1985). Refinement of a molecular model for lamprey hemoglobin from *Petromyzon marinus*. *J. Mol. Biol.* **184**, 147.

2LIV J. S. Sack, M. A. Saper, and F. A. Quiocho (1989). Periplasmic binding protein structure and function. Refined x-ray structures of the leucine/isoleucine/valine-binding protein and its complex with leucine. *J. Mol. Biol.* **206**, 171.

1LLC M. Buehner and H. J. Hecht. Structure determination of the allosteric L-lactate dehydrogenase from *Lactobacillus casei* at 3.0 Å resolution.

1LRD S. R. Jordan and C. O. Pabo (1988). Structure of the lambda complex at 2.5 Å resolution. Details of the repressor-operator interactions. *Science* **242**, 893.

1LRP D. H. Ohlendorf, W. F. Anderson, M. Lewis, C. O. Pabo, and B. W. Matthews (1983). Comparison of the structures of cro and λ repressor proteins from bacteriophage λ. *J. Mol. Biol.* **169**, 757.

1LYD D. R. Rose, J. Phipps, J. Michniewicz, G. I. Birnbaum, F. R. Ahmed, A. Muir, W. F. Anderson, and S. Narang (1988). Crystal structure of T4-lysozyme generated from synthetic coding DNA expressed in *Escherichia coli. Protein Eng.* **2**, 277.

1LYM S. T. Rao, J. Hogle, and M. Sundaralingam (1983). Studies of monoclinic hen egg white lysozyme. II. The refinement at 2.5 Å resolution—conformational variability between the two independent molecules. *Acta Crystallogr.* **C39**, 237.

2LYM, 3LYM C. E. Kundrot and F. M. Richards (1987). Crystal structure of hen egg-white lysozyme at a hydrostatic pressure of 1000 atmospheres. *J. Mol. Biol.* **193**, 157.

1LYZ, 2LYZ, 3LYZ, 4LYZ, 5LYZ, 6LYZ C. C. F. Blake, D. F. Koenig, G. A. Mair, A. C. T. North, D. C. Phillips, and V. R. Sarma (1965). Structure of hen egg-white lysozyme, a three-dimensional Fourier synthesis at 2 Å resolution. *Nature* **206**, 757.

7LYZ O. Herzberg and J. L. Sussman (1983). Protein model building by the use of a constrained-restrained least-squares procedure. *J. Appl. Crystallogr.* **16**, 144.

8LYZ C. R. Beddell, C. C. F. Blake, and S. J. Oatley (1975). An x-ray study of the structure and binding properties of iodine-inactivated lysozyme. *J. Mol. Biol.* **97**, 643.

9LYZ J. A. Kelly, A. R. Sielecki, B. D. Sykes, M. N. G. James, and D. C. Phillips (1979). X-ray crystallography of the binding of the bacterial cell wall trisaccharide NAM-NAG-NAM to lysozyme. *Nature* **282**, 875.

1LZ1 P. J. Artymiuk and C. C. F. Blake (1981). Refinement of human lysozyme at 1.5 Å resolution. Analysis of non-bonded and hydrogen-bond interactions. *J. Mol. Biol.* **152**, 737.

1LZ2 R. Sarma and R. Bott (1977). Crystallographic study of turkey egg-white lysozyme and its complex with a disaccharide. *J. Mol. Biol.* **113**, 555.

2LZ2 M. R. Parsons and S. E. V. Phillips. The three dimensional structure of turkey egg white lysozyme at 2.2 Å resolution.

1LZH, 2LZH P. J. Artymiuk, C. C. F. Blake, D. W. Rice, and K. S. Wilson (1982). The structures of the monoclinic and orthorhombic forms of hen egg-white lysozyme at 6 Å resolution. *Acta Crystallogr.* **B38**, 778.

2LZM, 3LZM L. H Weaver and B. W. Matthews (1987). Structure of bacteriophage T4 lysozyme refined at 1.7 Å resolution. *J. Mol. Biol.* **193**, 189

1LZT J. M. Hodsdon, G. M. Brown, L. C. Sieker, and L. H. Jensen (1990). Refinement of triclinic lysozyme. I. Fourier and least-squares methods. *Acta Crystallogr.* **B46**, 54.

2LZT M. Ramanadham, L. C. Sieker, and L. H. Jensen (1990) Refinement of triclinic lysozyme. II. The method of stereochemically restrained least squares. *Acta Crystallogr.* **B46**, 63.

1MB5 J. C. Hanson and B. P. Schoenborn (1981). Real space refinement of neutron diffraction data from sperm whale carbonmonoxymyoglobin. *J. Mol. Biol.* **153**, 117.

1MBA, 2MBA, 3MBA, 4MBA M. Bolognesi, S. Onesti, G. Gatti, A. Coda, P. Ascenzi, and M. Brunori (1989). *Aplysia limacina* myoglobin. Crystallographic analysis at 1.6 Å resolution. *J. Mol. Biol.* **205**, 529.

1MBC J. Kuriyan, S. Wilz, M. Karplus, and G. A. Petsko (1986). X-ray structure and refinement of carbon-monoxy (Fe II)- myoglobin at 1.5 Å resolution. *J. Mol. Biol.* **192**, 133.

1MBD S. E. V. Phillips and B. P. Schoenborn (1981). Neutron diffraction reveals oxygen-histidine hydrogen bond in oxymyoglobin. *Nature* **292**, 81.

1MBN H. C. Watson (1969). The stereochemistry of the protein myoglobin. *Prog. Stereochem.* **4**, 299.

4MBN, 5MBN T. Takano (1984). Refinement of myoglobin and cytochrome c. *Methods and applications in crystallographic computing*, p. 262 (S.R. Hall and T. Ashida, Ed.) Oxford University Press, Oxford, England.

1MBO S. E. V. Phillips (1980). Structure and refinement of oxymyoglobin at 1.6 Å resolution. *J. Mol. Biol.* **142**, 531.

1MBS H. Scouloudi and E. N. Baker (1978). X-ray crystallographic studies of seal myoglobin. The molecule at 2.5 Å resolution. *J. Mol. Biol.* **126**, 637.

1MCG M. Schiffer, F. J. Stevens, F. A. Westholm, S. S. Kim, and R. D. Carlson (1982). Small-angle neutron scattering study of Bence-Jones protein MCG. Comparison of structures in solution and in crystal. *Biochemistry* **21**, 2874.

1MCP Y. Satow, G. H. Cohen, E. A. Padlan, and D. R. Davies (1987). Phosphocholine binding immunoglobulin Fab McPC603. An x-ray diffraction study at 2.7 Å. *J. Mol. Biol.* **190**, 593.

2MCP E. A. Padlan, G. H. Cohen, and D. R. Davies. Refined crystal structure of the McPC603 Fab-phosphocholine complex at 3.1 Å resolution.

4MDH J. J. Birktoft, G. Rhodes, and L. J. Banaszak (1989). Refined crystal structure of cytoplasmic malate dehydrogenase at 2.5 Å resolution. *Biochemistry* **28**, 6065.

2MEV S. Krishnaswamy and M. G. Rossmann (1990). Structural refinement and analysis of mengo virus. *J. Mol. Biol.* **211**, 803.

2MHB R. C. Ladner, E. G. Heidner, and M. F. Perutz (1977). The structure of horse methaemoglobin at 2.0 Å resolution. *J. Mol. Biol.* **114**, 385.

2MHR S. Sheriff, W. A. Hendrickson, and J. L. Smith (1987). Structure of myohemerythrin in the azidomet state at 1.7/1.3 Å resolution. *J. Mol. Biol.* **197**, 273.

1MLP A. D. McLachlan (1978). The double helix coiled coil structure of murein lipoprotein from *Escherichia coli*. *J. Mol. Biol.* **121**, 493.

1MLT T. C. Terwilliger and D. Eisenberg (1982). The structure of melittin. I. Structure determination and partial refinement. *J. Biol. Chem.* **257**, 6010.

1MON C. Ogata, M. Hatada, G. Tomlinson, W.-C. Shin, and S.-H. Kim (1987). Crystal structure of the intensely sweet protein monellin. *Nature* **328**, 739.

1NTP A. A. Kossiakoff (1984). Use of the neutron diffraction-H/D exchange technique to determine the conformational dynamics of trypsin. *Basic Life Sci.* **27**, 281.

1NXB D. Tsernoglou, G. A. Petsko, and R. A. Hudson (1978). Structure and function of snake venom curarimimetic neurotoxins. *Mol. Pharmacol.* **14**, 710.

1OVO E. Papamokos, E. Weber, W. Bode, R. Huber, M. W. Empie, I. Kato, and M. Laskowski, Jr (1982). Crystallographic refinement of Japanese quail ovomucoid, a Kazal-type inhibitor, and model building studies of complexes with serine proteases. *J. Mol. Biol.* **158**, 515.

2OVO W. Bode, O. Epp, R. Huber, M. Laskowski, Jr., and W. Ardelt (1985). The crystal and molecular structure of the third domain of silver pheasant ovomucoid (OMSVP3). *Eur. J. Biochem.* **147**, 387.

1P01 R. Bone, A. B. Shenvi, C. A. Kettner, and D. A. Agard (1987). Serine protease mechanism. Structure of an inhibitory complex of α-lytic protease and a tightly bound peptide boronic acid. *Biochemistry* **26**, 7609.

1P02, 1P03, 1P04, 1P05, 1P06 R. Bone, D. Frank, C. Kettner, and D. A. Agard (1989). Structure analysis of specificity. α-lytic protease complexes with analogues of reaction intermediates. *Biochemistry* **28**, 7600.

1P07, 1P08, 1P09, 1P10 R. Bone, J. L. Silen, and D. A. Agard (1989). Structural plasticity broadens the specificity of an engineered protease. *Nature* **339**, 191.

2P21, 3P21 L. Tong, M. V. Milburn, A. M. de Vos, and S.-H. Kim (1989). Structure of *ras* protein. *Science* **245**, 244.

1P2P B. W. Dijkstra, R. Renetseder, K. H. Kalk, W. G. J. Hol, and J. Drenth (1983). Structure of porcine pancreatic phospholipase A_2 at 2.6 Å resolution and comparison with bovine phospholipase A_2. *J. Mol. Biol.* **168**, 163.

3P2P O. P. Kuipers, M. M. G. M. Thunnissen, P. de Geus, B. W. Dijkstra, H. M. Verheij, and G. H. de Haas (1989). Enhanced activity and altered specificity of Phospholipase A$_2$ by deletion of a surface loop. *Science* **244**, 82.

2PAB C. C. F. Blake, M. J. Geisow, S. J. Oatley, B. Rerat, and C. Rerat (1978). Structure of prealbumin, secondary, tertiary and quaternary interactions determined by Fourier refinement at 1.8 Å. *J. Mol. Biol.* **121**, 339.

1PAD, 2PAD, 4PAD, 5PAD, 6PAD J. Drenth, K. H. Kalk, and H. M. Swen (1976). Binding of chloromethyl ketone substrate analogues to crystalline papain. *Biochemistry* **15**, 3731.

9PAP I. G. Kamphuis, K. H. Kalk, M. B. A. Swarte, and J. Drenth (1984). Structure of papain refined at 1.65 Å resolution. *J. Mol. Biol.* **179**, 233.

1PAZ K. Petratos, Z. Dauter, and K. S. Wilson (1988). The refinement of the structure of pseudoazurin from *Alcaligenes faecalis s-6* at 1.55 Å resolution. *Acta Crystallogr.* **B44**, 628.

2PAZ E. T. Adman, S. Turley, R. Bramson, K. Petratos, D. Banner, D. Tsernoglou, T. Beppu, and H. Watanabe (1989). A 2.0-Å structure of the blue copper protein (cupredoxin) from *Alcaligenes faecalis s-6*. *J. Biol. Chem.* **264**, 87.

1PCY J. M. Guss and H. C. Freeman (1983). Structure of oxidized poplar plastocyanin at 1.6 Å resolution. *J. Mol. Biol.* **169**, 521.

2PCY T. P. J. Garrett, D. J. Clingeleffer, J. M. Guss, S. J. Rogers, and H. C. Freeman (1984). The crystal structure of poplar apoplastocyanin at 1.8-Å resolution. The geometry of the copper-binding site is created by the polypeptide. *J. Biol. Chem.* **259**, 2822.

3PCY W. B. Church, J. M. Guss, J. J. Potter, and H. C. Freeman (1986). The crystal structure of mercury-substituted poplar plastocyanin at 1.9-Å resolution. *J. Biol. Chem.* **261**, 234.

4PCY, 5PCY, 6PCY J. M. Guss, P. R. Harrowell, M. Murata, V. A. Norris, and H. C. Freeman (1986). Crystal structure analyses of reduced (Cu^I) poplar plastocyanin at six pH values. *J. Mol. Biol.* **192**, 361.

1PEP N. S. Andreeva, A. A. Fedorov, A. E. Gutschina, R. R. Riskulov, N. E. Schutzkever, and M. G. Safro (1978). X-ray crystallographic studies of pepsin. Conformation of the main chain of the enzyme. *Mol. Biol. (Moscow)* **12**, 922.

2PEP J. B. Cooper, G. Khan, G. Taylor, I. J. Tickle, and T. L. Blundell (1990). Three dimensional structure of the hexagonal crystal form of porcine pepsin at 2.3 Å resolution. *J. Mol. Biol.* **214,** 199.

3PEP C. Abad-Zapatero, T. J. Rydel, and J. Erickson. Revised 2.3 Å structure of porcine pepsin. Evidence for a flexible subdomain.

4PEP A. R. Sielecki, A. A. Fedorov, A. Boodhoo, N. S. Andreeva, and M. N. G. James (1990). The molecular and crystal structures of monoclinic porcine pepsin refined at 1.8 Å resolution. *J. Mol. Biol.* **214**, 143.

1PFC S. H. Bryant, L. M. Amzel, R. P. Phizackerley, and R. J. Poljak (1985). Molecular-replacement structure of guinea pig IgG1 pFc′ refined at 3.1 Å resolution. *Acta Crystallogr.* **B41**, 362.

1PFK Y. Shirakihara and P. R. Evans (1988). Crystal structure of the complex of phosphofructokinase from *Escherichia coli* with its reaction products. *J. Mol. Biol.* **204**, 973.

Appendix 1

2PFK W. R. Rypniewski and P. R. Evans (1989). Crystal structure of unliganded phospho-fructokinase from *Escherichia coli*. *J. Mol. Biol.* **207**, 805.

3PFK, 4PFK P. R. Evans, G. W. Farrants, and P. J. Hudson (1981). Phosphofructo-kinase. Structure and control. *Philos. Trans. R. Soc. London* **293**, 53.

5PFK P. R. Evans, G. W. Farrants and M. C. Lawrence (1986). Crystallographic structure of allosterically inhibited phosphofructokinase at 7 Å resolution. *J. Mol. Biol.* **191**, 713.

1PGI P. J. Shaw and H. Muirhead (1977). Crystallographic structure analysis of glucose 6-phosphate isomerase at 3.5 Å resolution. *J. Mol. Biol.* **109**, 475.

2PGK D. W. Rice (1981). The use of phase combination in the refinement of phosphoglycerate kinase at 2.5 Å resolution. *Acta Crystallogr.* **A37**, 491.

3PGK T. N. Bryant, P. J. Shaw, N. P. Walker, P. L. Wendell, and H. C. Watson. The structure of yeast phosphoglycerate kinase at 0.25 nm resolution.

3PGM S. I. Winn, J. Warwicker, and H. C. Watson. The structure of yeast phosphoglycerate mutase at 0.28 nm resolution.

1PHH H. A. Schreuder, J. M. van der Laan, W. G. J. Hol, and J. Drenth (1988). Crystal structure of *p*-hydroxybenzoate hydroxylase complexed with its reaction product 3,4-dihydroxybenzoate. *J. Mol. Biol.* **199**, 637.

1PHY D. E. McRee, J. A. Tainer, T. E. Meyer, J. van Beeumen, M. A. Cusanovich, and E. D. Getzoff (1989). Crystallographic structure of a photoreceptor protein at 2.4 Å resolution. *Proc. Nat. Acad. Sci. USA* **86**, 6533.

2PKA W. Bode, Z. Chen. K. Bartels, C. Kutzbach, G. Schmidt-Kastner, and H. Bartunik (1983). Refined 2 Å x-ray crystal structure of porcine pancreatic kallikrein a, a specific trypsin-like serine proteinase. Crystallization, structure determination, crystallographic refinement, structure and its comparison with bovine trypsin. *J. Mol. Biol.* **164**, 237.

2PLV D. J. Filman, R. Syed, M. Chow, A. J. Macadam, P. D. Minor, J. M. Hogle (1989). Structural factors that control conformational transitions and serotype specificity in type 3 poliovirus. *EMBO Journal* **8**, 1567.

1PMB S. J. Smerdon, T. J. Oldfield, E. J. Dodson, G. G. Dodson, R. E. Hubbard, and A. J. Wilkinson (1990). The determination of the crystal structure of recombinant pig myoglobin by molecular replacement and its refinement. *Acta Crystallogr.* **B46**, 370.

1PP2 S. Brunie, J. Bolin, D. Gewirth, and P. B. Sigler (1985). The refined crystal structure of dimeric phospholipase A_2 at 2.5 Å. Access to a shielded catalytic center. *J. Biol. Chem.* **260**, 9742.

1PPD J. P. Priestle, G. C. Ford, M. Glor, E. L. Mehler, J. D. G. Smit, C. Thaller, and J. N. Jansonius (1984). Restrained least-squares refinement of the sulfhydryl protease papain to 2.0 Å. *Acta Crystallogr.* **A40**, 17.

1PPT T. L. Blundell, J. E. Pitts, I. J. Tickle, S. P. Wood, and C.-W. Wu (1981). X-ray analysis (1.4-Å resolution) of avian pancreatic polypeptide. Small globular protein hormone. *Proc. Nat. Acad. Sci. USA* **78**, 4175.

1PRC J. Deisenhofer, O. Epp, I. Sinning, and H. Michel. Crystallographic refinement at 2.3 Å resolution and refined model of the photosynthetic reaction center from *Rhodopseudomonas viridis*.

2PRK C. Betzel, G. P. Pal, and W. Saenger. Synchrotron x-ray data collection and restrained least-squares refinement of the crystal structure of proteinase K at 1.5 Å resolution.

1PSG J. A. Hartsuck and S. J. Remington. The structure of porcine pepsinogen at 1.65 Å resolution.

3PTB, 2PTC M. Marquart, J. Walter, J. Deisenhofer, W. Bode, and R. Huber (1983). The geometry of the reactive site and of the peptide groups in trypsin, trypsinogen and its complexes with inhibitors. *Acta Crystallogr.* **B39**, 480.

1PTE J. A. Kelly, J. R. Knox, P. C. Moews, G. J. Hite, J. B. Bartolone, H. Zhao, B. Joris, J.-M. Frère, and J.-M. Ghuysen (1985). 2.8-Å structure of penicillin-sensitive D-alanyl carboxypeptidase-transpeptidase from *Streptomyces r61* and complexes with β-lactams. *J. Biol. Chem.* **260**, 6449.

4PTI M. Marquart, J. Walter, J. Deisenhofer, W. Bode, and R. Huber (1983). The geometry of the reactive site and of the peptide groups in trypsin, trypsinogen and its complexes with inhibitors. *Acta Crystallogr.* **B39**, 480.

5PTI A. Wlodawer, J. Walter, R. Huber, and L. Sjölin (1984). Structure of bovine pancreatic trypsin inhibitor. Results of joint neutron and x-ray refinement of crystal form II. *J. Mol. Biol.* **180**, 301.

6PTI A. Wlodawer, J. Nachman, G. L. Gilliland, W. Gallagher, and C. Woodward (1987). Structure of form III crystals of bovine pancreatic trypsin inhibitor. *J. Mol. Biol.* **198**, 469.

2PTN, 3PTN J. Walter, W. Steigemann, T. P. Singh, H. Bartunik, W. Bode, and R. Huber (1982). On the disordered activation domain in trypsinogen. Chemical labelling and low-temperature crystallography. *Acta Crystallogr.* **B38**, 1462.

4PTP J. L. Chambers and R. M. Stroud (1979). The accuracy of refined protein structures. Comparison of two independently refined models of bovine trypsin. *Acta Crystallogr.* **B35**, 1861.

1PYK D. I. Stuart, M. Levine, H. Muirhead, and D. K. Stammers (1979). Crystal structure of cat muscle pyruvate kinase at a resolution of 2.6 Å. *J. Mol. Biol.* **134**, 109.

1PYP E. G. Arutiunian, S. S. Terzian, A. A. Voronova, I. P. Kuranova, E. A. Smirnova, B. K. Vainshtein, W. E. Hone, and G. Hansen (1981). X-ray diffraction study of inorganic pyrophosphatase from baker's yeast at 3 Å resolution (Russian). *Dokl. Akad. Nauk SSSR* **258**, 1481.

2R04, 2R06, 2R07, 1R08 J. Badger, I. Minor, M. A. Oliveira, T. J. Smith, and M. G. Rossmann (1989). Structural analysis of antiviral agents that interact with the capsid of human rhinoviruses. *Proteins: Struct., Funct., Genet.* **6**, 1.

1R69 A. Mondragon, S. Subbiah, S. C. Almo, M. Drottar, and S. C. Harrison (1989). Structure of the amino-terminal domain of phage 434 repressor at 2.0 Å resolution. *J. Mol. Biol.* **205**, 189.

1RBB R. L. Williams, S. M. Greene, and A. McPherson (1987). The crystal structure of ribonuclease B at 2.5-Å resolution. *J. Biol. Chem.* **262**, 16020.

1RDG M. Frey, L. Sieker, F. Payan, R. Haser, M. Bruschi, G. Pepe, and J. LeGall (1987). Rubredoxin from *Desulfovibrio gigas*. A molecular model of the oxidized form at 1.4 Å resolution. *J. Mol. Biol.* **197**, 525.

1REI O. Epp, E. E. Lattman, M. Schiffer, R. Huber, and W. Palm (1975). The molecular structure of a dimer composed of the variable portions of the Bence-Jones protein REI refined at 2.0 Å resolution. *Biochemistry* **14**, 4943.

1RHD J. H. Ploegman, G. Drent, K. H. Kalk, and W. G. J. Hol (1978). Structure of bovine liver rhodanese. I. Structure determination at 2.5 Å resolution and a comparison of the conformation and sequence of its two domains. *J. Mol. Biol.* **123**, 557.

2RHE W. Furey, Jr., B. C. Wang, C. S. Yoo, and M. Sax (1983). Structure of a novel Bence-Jones protein (RHE) fragment at 1.6 Å resolution. *J. Mol. Biol.* **167**, 661.

4RHV E. Arnold and M. G. Rossmann (1990). Analysis of the structure of a common cold virus, human rhinovirus 14, refined at a resolution of 3.0 Å. *J. Mol. Biol.* **211**, 763.

1RLX, 2RLX, 3RLX, 4RLX N. Isaacs, R. James, H. Niall, G. Bryant-Greenwood, G. Dodson, A. Evans, and A. C. T. North (1978). Relaxin and its structural relationship to insulin. *Nature* **271**, 278.

2RM2 J. Badger, I. Minor, M. A. Oliveira, T. J. Smith, and M. G. Rossmann (1989). Structural analysis of antiviral agents that interact with the capsid of human rhinoviruses. *Proteins: Struct., Funct., Genet.* **6**, 1.

1RMU, 2RMU J. Badger, S. Krishnaswamy, M. J. Kremer, M. A. Oliveira, M. G. Rossmann, B. A. Heinz, R. R. Rueckert, F. J. Dutko, and M. A. Kinlay (1989). Three-dimensional structures of drug-resistant mutants of human rhinovirus 14. *J. Mol. Biol.* **207**, 163.

1RN3 N. Borkakoti, D. S. Moss, and R. A. Palmer (1982). Ribonuclease-A. Least-squares refinement of the structure at 1.45 Å resolution. *Acta Crystallogr.* **B38**, 2210 (1982).

1RNS R. J. Fletterick and H. W. Wyckoff (1975). Preliminary refinement of protein coordinates in real space. *Acta Crystallogr.* **A31**, 698.

1RNT R. Arni, U. Heinemann, M. Maslowska, R. Tokuoka, and W. Saenger (1987). Restrained least-squares refinement of the crystal structure of the ribonuclease T_1*2'-guanylic acid complex at 1.9 Å resolution. *Acta Crystallogr.* **B43**, 549.

2RNT J. Koepke, M. Maslowska, U. Heinemann, and W. Saenger (1985). Three-dimensional structure of ribonuclease T_1 complexed with guanylyl-2',5'-guanosine at 1.8 Å resolution. *J. Mol. Biol.* **206**, 475.

3RNT D. Kostrewa, H.-W. Choe, U. Heinemann, and W. Saenger (1989). Crystal structure of guanosine-free ribonuclease T_1, complexed with vanadate(V), suggests conformational change upon substrate binding. *Biochemistry* **28**, 7592.

3RP2 S. J. Remington, R. G. Woodbury, R. A. Reynolds, B. W. Matthews, and H. Neurath (1988). The structure of rat mast cell protease II at 1.9-Å resolution. *Biochemistry* **27**, 8097.

2RR1, 2RS1, 2RS3, 2RS5 J. Badger, I. Minor, M. A. Oliveira, T. J. Smith, and M. G. Rossmann (1989). Structural analysis of antiviral agents that interact with the capsid of human rhinoviruses. *Proteins: Struct., Funct., Genet.* **6**, 1.

5RSA A. Wlodawer, N. Borkakoti, D. S. Moss, and B. Howlin (1986). Comparison of two independently refined models of ribonuclease-A. *Acta Crystallogr.* **B42**, 379.

6RSA B. Borah, C.-W. Chen, W. Egan, M. Miller, A. Wlodawer, and J. S. Cohen (1985). Nuclear magnetic resonance and neutron diffraction studies of the complex of ribonuclease A with uridine vanadate, a transition-state analogue. *Biochemistry* **24**, 2058.

Appendix 1

7RSA A. Wlodawer, L. A. Svensson, L. Sjölin, and G. L. Gilliland (1988). Structure of phosphate-free ribonuclease A refined at 1.26 Å. *Biochemistry* **27**, 2705.

1RSM P. C. Weber, S. Sheriff, D. H. Ohlendorf, B. C. Finzel, and F. R. Salemme (1985). The 2-Å resolution structure of a thermostable ribonuclease A chemically cross-linked between lysine residues 7 and 41. *Proc. Nat. Acad. Sci. USA* **82**, 8473.

2RSP M. Jaskólski, M. Miller, J. K. M. Rao, J. Leis, and A. Wlodawer (1990). Structure of the aspartic protease from Rous sarcoma retrovirus refined at 2 Å resolution. *Biochemistry* **29**, 5889.

2RUB I. Andersson, S. Knight, G. Schneider, Y. Lindqvist, T. Lundqvist, C.-I. Brändén, and G. H. Lorimer (1985). Crystal structure of the active site of ribulose-bisphosphate carboxylase. *Nature* **337**, 229.

3RXN E. T. Adman and L. H. Jensen (1979). Progress on refinement of rubredoxin (*D. vulgaris*) at 1.5 Å. *Am. Cryst. Assoc., Abstr. Papers* **6**, 65.

4RXN K. D. Watenpaugh, L. C. Sieker, and L. H. Jensen (1980). Crystallographic refinement of rubredoxin at 1.2 Å resolution. *J. Mol. Biol.* **138**, 615.

5RXN K. D. Watenpaugh. Combined crystallographic refinement and energy minimization of rubredoxin at 1.2 Å resolution.

1SBC D. J. Neidhart and G. A. Petsko (1988). The refined crystal structure of subtilisin Carlsberg at 2.5 Å resolution. *Protein Engineering* **2**, 271.

1SBT R. A. Alden, J. J. Birktoft, J. Kraut, J. D. Robertus, and C. S. Wright (1971). Atomic coordinates for subtilisin BPN (or Novo). *Biochem. Biophys. Res. Comm.* **45**, 337.

2SBT J. Drenth, W. G. J. Hol, J. N. Jansonius, and R. Koekoek (1972). A comparison of the three-dimensional structures of subtilisin BPN' and subtilisin novo. *Cold Spr. Harb. Symp. Quant. Biol.* **36**, 107.

4SBV A. M. Silva and M. G. Rossmann (1975). The refinement of southern bean virus in reciprocal space. *Acta Crystallogr.* **B41**, 147.

2SEC C. A. McPhalen and M. N. G. James (1988). Structural comparison of two serine proteinase-protein inhibitor complexes. Eglin-c-subtilisin carlsberg and CI-2-subtilisin novo. *Biochemistry* **27**, 6582.

2SGA J. Moult, F. Sussman, and M. N. G. James (1985). Electron density calculations as an extension of protein structure refinement. *Streptomyces griseus* protease A at 1.5 Å resolution. *J. Mol. Biol.* **182**, 555.

3SGB R. J. Read, M. Fujinaga, A. R. Sielecki, and M. N. G. James (1983). Structure of the complex of *Streptomyces griseus* protease b and the third domain of the turkey ovomucoid inhibitor at 1.8 Å resolution. *Biochemistry* **22**, 4420.

1SGC L. T. J. Delbaere and G. D. Brayer (1985). The 1.8 Å structure of the complex between chymostatin and *Streptomyces griseus* protease a. A model for serine protease catalytic tetrahedral intermediates. *J. Mol. Biol.* **183**, 89.

1SGT R. J. Read and M. N. G. James (1988). Refined crystal structure of *Streptomyces griseus* trypsin at 1.7 Å resolution. *J. Mol. Biol.* **200**, 523.

1SIC S. Hirono, H. Akagawa, Y. Mitsui, and Y. Iitaka (1984). Crystal structure at 2.6 Å resolution of the complex of subtilisin BPN' with *Streptomyces* subtilisin inhibitor. *J. Mol. Biol.* **178**, 389.

Appendix 1

1SN3 R. J. Almassy, J. C. Fontecilla-Camps, F. L. Suddath, and C. E. Bugg (1983). Structure of variant-3 scorpion neurotoxin from *Centruroides sculpturatus ewing*, refined at 1.8 Å resolution. *J. Mol. Biol.* **170**, 497.

2SNI C. A. McPhalen and M. N. G. James (1988). Structural comparison of two serine proteinase-protein inhibitor complexes. Eglin-c-subtilisin carlsberg and CI-2-subtilisin novo. *Biochemistry* **27**, 6582.

2SNS F. A. Cotton, E. E. Hazen, Jr, and M. J. Legg (1979). Staphylococcal nuclease. Proposed mechanism of action based on structure of enzyme-thymidine-3′,5′-biphosphate-calcium ion complex at 1.5-Å resolution. *Proc. Nat. Acad. Sci. USA* **76**, 2551.

2SOD J. A. Tainer, E. D. Getzoff, K. M. Beem, J. S. Richardson, and D. C. Richardson (1982). Determination and analysis of the 2 Å structure of copper, zinc superoxide dismutase. *J. Mol. Biol.* **160**, 181.

1SRX A. Holmgren, B.-O. Söderberg, H. Eklund, and C.-I. Brändén (1975). Three-dimensional structure of *Escherichia coli* thioredoxin-S$_2$ to 2.8 Å resolution. *Proc. Nat. Acad. Sci. USA* **72**, 2305.

2SSI Y. Satow, Y. Watanabe, and Y. Mitsui (1980). Solvent accessibility and microenvironment in a bacterial protein proteinase inhibitor SSI (*Streptomyces* subtilisin inhibitor). *J. Biochem. (Tokyo)* **88**, 1739.

2STV L. Liljas and B. Strandberg (1984). Structure of satellite tobacco necrosis virus after crystallographic refinement at 2.5 Å resolution. *J. Mol. Biol.* **177**, 735

2TAA Y. Matsuura, M. Kusunoki, W. Harada, and M. Kakudo (1984). Structure and possible catalytic residues of taka-amylase a. *J. Biochem. (Tokyo)* **95**, 697.

2TBV A. J. Olsen, G. Bricogne, and S. C. Harrison (1983). Structure of tomato bushy stunt virus. IV. The virus particle at 2.9 Å resolution. *J. Mol. Biol.* **171**, 61.

1TEC P. Gros, M. Fujinaga, B. W. Dijkstra, K. H. Kalk, and W. G. J. Hol (1989). Crystallographic refinement by incorporation of molecular dynamics. The thermostable serine protease thermitase complexed with eglin-c. *Acta Crystallogr.* **B45**, 488.

2TGA J. Walter, W. Steigemann, T. P. Singh, H. Bartunik, W. Bode, and R. Huber (1982). On the disordered activation domain in trypsinogen. Chemical labelling and low-temperature crystallography. *Acta Crystallogr.* **B38**, 1462.

1TGB H. Fehlhammer, W. Bode, and R. Huber (1977). Crystal structure of bovine trypsinogen at 1.8 Å resolution. II. Crystallographic refinement, refined crystal structure and comparison with bovine trypsin. *J. Mol. Biol.* **111**, 415.

1TGC J. Walter, W. Steigemann, T. P. Singh, H. Bartunik, and R. Huber (1982). On the disordered activation domain in trypsinogen. Chemical labelling and low-temperature crystallography. *Acta Crystallogr.* **B38**, 1462.

2TGD M. O. Jones and R. M. Stroud. Lack of the transition state stabilization site is a factor in the inactivity of trypsinogen, a serine protease zymogen. Structure of DFP inhibited bovine trypsinogen at 2.1 Å resolution.

1TGN A. A. Kossiakoff, J. L. Chambers, L. M. Kay, and R. M. Stroud (1977). Structure of bovine trypsinogen at 1.9 Å resolution. *Biochemistry* **16**, 654.

2TGP M. Marquart, J. Walter, J. Deisenhofer, W. Bode, and R. Huber (1983). The geometry of the reactive site and of the peptide groups in trypsin, trypsinogen and its complexes with inhibitors. *Acta Crystallogr.* **B39**, 480.

1TGS M. Bolognesi, G. Gatti, E. Menegatti, M. Guarneri, M. Marquart, E. Papamokos, and R. Huber (1982). Three-dimensional structure of the complex between pancreatic secretory inhibitor (Kazal type) and trypsinogen at 1.8 Å resolution. Structure solution, crystallographic refinement and preliminary structural interpretation. *J. Mol. Biol.* **162**, 839.

1TGT, 2TGT J. Walter, W. Steigemann, T. P. Singh, H. Bartunik, W. Bode, and R. Huber (1982). On the disordered activation domain in trypsinogen. Chemical labelling and low-temperature crystallography. *Acta Crystallogr.* **B38**, 1462.

1THI A. M. de Vos, M. Hatada, H. van der Wel, H. Krabbendam, A. F. Peerdeman, and S.-H. Kim (1985). Three-dimensional structure of thaumatin I, an intensely sweet protein. *Proc. Nat. Acad. Sci. USA* **82**, 1406.

1TIM D. W. Banner, A. C. Bloomer, G. A. Petsko, D. C. Phillips, C. I. Pogson, I. A. Wilson, P. H. Corran, A. J. Furth, J. D. Milman, R. E. Offord, J. D. Priddle, and S. G. Waley (1975). Structure of chicken muscle triose phosphate isomerase determined crystallographically at 2.5 Å resolution using amino acid sequence data. *Nature* **255**, 609.

1TLD H. D. Bartunik, L. J. Summers, and H. H. Bartsch (1989). Crystal structure of bovine β-trypsin at 1.5 Å resolution in a crystal form with low molecular packing density. Active site geometry, ion pairs and solvent structure. *J. Mol. Biol.* **210**, 813.

3TLN M. A. Holmes and B. W. Matthews (1982). Structure of thermolysin refined at 1.6 Å resolution. *J. Mol. Biol.* **160**, 623.

4TLN, 5TLN, 7TLN M. A. Holmes and B. W. Matthews (1981). Binding of hydroxamic acid inhibitors to crystalline thermolysin suggests a pentacoordinate zinc intermediate in catalysis. *Biochemistry* **20**, 6912.

1TLP D. E. Tronrud, A. F. Monzingo, and B. W. Matthews (1986). Crystallographic structural analysis of phosphoramidates as inhibitors and transition-state analogs of thermolysin. *Eur. J. Biochem.* **157**, 261.

2TMA G. N. Phillips, Jr (1986). Tropomyosin crystal structure and muscle regulation. Appendix. Construction of an atomic model for tropomyosin and implications for interactions with actin. *J. Mol. Biol.* **192**, 128.

1TMN A. F. Monzingo and B. W. Matthews (1984). Binding of N-carboxymethyl dipeptide inhibitors to thermolysin determined by x-ray crystallography. A novel class of transition-state analogues for zinc peptidases. *Biochemistry* **23**, 5724.

2TMN D. E. Tronrud, A. F. Monzingo, and B. W. Matthews (1986). Crystallographic structural analysis of phosphoramidates as inhibitors and transition-state analogs of thermolysin. *Eur. J. Biochem.* **157**, 261.

3TMN H. M. Holden and B. W. Matthews (1988). The binding of L-valyl-L-tryptophan to crystalline thermolysin illustrates the mode of interaction of a product of peptide hydrolysis. *J. Biol. Chem.* **263**, 3256.

4TMN, 5TMN H. M. Holden, D. E. Tronrud, A. F. Monzingo, L. H. Weaver, and B. W. Matthews (1987). Slow- and fast-binding inhibitors of thermolysin display different modes of binding. Crystallographic analysis of extended phosphonamidate transition-state analogues. *Biochemistry* **26**, 8542.

6TMN D. E. Tronrud, H. M. Holden, and B. W. Matthews (1987). Structures of two thermolysin-inhibitor complexes that differ by a single hydrogen bond. *Science* **235**, 571.

7TMN H. M. Holden, D. E. Tronrud, A. F. Monzingo, L. H. Weaver, and B. W. Matthews (1987). Slow- and fast-binding inhibitors of thermolysin display different modes of binding. Crystallographic analysis of extended phosphonamidate transition-state analogues. *Biochemistry* **26**, 8542.

2TMV K. Namba, R. Pattanayek, and G. Stubbs (1989). Visualization of protein-nucleic acid interactions in a virus. Refined structure of intact tobacco mosaic virus at 2.9 Å resolution by x-ray fiber diffraction. *J. Mol. Biol.* **208**, 307.

1TN1, 1TN2 R. S. Brown, J. C. Dewan, and A. Klug (1985). Crystallographic and biochemical investigation of the lead(II)-catalyzed hydrolysis of yeast phenylalanine tRNA. *Biochemistry* **24**, 4785.

4TNA B. E. Hingerty, R. S. Brown, and A. Jack (1978). Further refinement of the structure of yeast tRNA^{phe}. *J. Mol. Biol.* **124**, 523.

6TNA J. L. Sussman, S. R. Holbrook, R. W. Warrant, G. M. Church, and S.-H. Kim (1978). Crystal structure of yeast phenylalanine tRNA. I. Crystallographic refinement. *J. Mol. Biol.* **123**, 607.

1TNC R. H. Kretsinger and C. D. Barry (1975). The predicted structure of the calcium-binding component of troponin. *Biochim. Biophys. Acta* **405**, 40.

4TNC K. A. Satyshur, S. T. Rao, D. Pyzalska, W. Drendel, M. Greaser, and M. Sundaralingam (1988). Refined structure of chicken skeletal muscle troponin c in the two-calcium state at 2-Å resolution. *J. Biol. Chem.* **263**, 1628.

5TNC O. Herzberg and M. N. G. James (1988). Refined crystal structure of troponin c from turkey skeletal muscle at 2.0 Å resolution. *J. Mol. Biol.* **203**, 761.

1TNF M. J. Eck and S. R. Sprang (1989). The structure of tumor necrosis factor-α at 2.6 Å resolution. Implications for receptor binding. *J. Biol. Chem.* **264**, 17595.

1TON M. Fujinaga and M. N. G. James (1987). Rat submaxillary gland serine protease, tonin. Structure solution and refinement at 1.8 Å resolution. *J. Mol. Biol.* **195**, 373.

1TPA M. Marquart, J. Walter, J. Deisenhofer, W. Bode, and R. Huber (1983). The geometry of the reactive site and of the peptide groups in trypsin, trypsinogen and its complexes with inhibitors. *Acta Crystallogr.* **B39**, 480.

2TPI J. Walter, W. Steigemann, T. P. Singh, H. Bartunik, W. Bode, and R. Huber (1982). On the disordered activation domain in trypsinogen. Chemical labelling and low-temperature crystallography. *Acta Crystallogr.* **B38**, 1462.

3TPI, 1TPO, 1TPP M. Marquart, J. Walter, J. Deisenhofer, W. Bode, and R. Huber (1983). The geometry of the reactive site and of the peptide groups in trypsin, trypsinogen and its complexes with inhibitors. *Acta Crystallogr.* **39**, 480.

4TPI W. Bode, J. Walter, R. Huber, H. R. Wenzel, and H. Tschesche (1984). The refined 2.2-Å (0.22-nm) x-ray crystal structure of the ternary complex formed by bovine trypsinogen, valine-valine and the arg$_{15}$ analogue of bovine pancreatic trypsin inhibitor. *Eur. J. Biochem.* **144**, 185.

1TRA E. Westhof and M. Sundaralingam (1986). Restrained refinement of the monoclinic form of yeast phenylalanine transfer RNA. Temperature factors and dynamics, coordinated waters, and base-pair propeller twist angles. *Biochemistry* **25**, 4868.

2TRA, 3TRA, 4TRA E. Westhof, P. Dumas, and D. Moras (1988). Restrained refinement of two crystalline forms of yeast aspartic acid and phenylalanine transfer RNA crystals. *Acta Crystallogr.* **A44**, 112.

1TRM, 2TRM S. Sprang, T. Standing, R. J. Fletterick, R. M. Stroud, J. Finer-Moore, N.-H. Xuong, R. Hamlin, W. J. Rutter, and C. S. Craik (1987). The three-dimensional structure of asn$_{102}$ mutant of trypsin. Role of asp$_{102}$ in serine protease catalysis. *Science* **237**, 905.

2TS1, 3TS1 P. Brick, T. N. Bhat, and D. M. Blow (1989). Structural of tyrosyl-tRNA synthetase refined at 2.3 Å resolution. Interaction of the enzyme with the tyrosyl adenylate intermediate. *J. Mol. Biol.* **208**, 83.

4TS1 P. Brick and D. M. Blow (1987). Crystal structure of a deletion mutant of a tyrosyl-tRNA synthetase complexed with tyrosine. *J. Mol. Biol.* **194**, 287.

1UBQ S. Vijay-Kumar, C. E. Bugg, and W. J. Cook (1987). Structure of ubiquitin refined at 1.8 Å resolution. *J. Mol. Biol.* **194**, 531.

1UTG I. Morize, E. Surcouf, M. C. Vaney, Y. Epelboin, M. Buehner, F. Fridlansky, E. Milgrom, and J. P. Mornon (1987). Refinement of the C222$_1$ crystal form of oxidized uteroglobin at 1.34 Å resolution. *J. Mol. Biol.* **194**, 725.

2UTG R. Bally and J. Delettre (1989). Structure and refinement of the oxidized P2$_1$ form of uteroglobin at 1.64 Å resolution. *J. Mol. Biol.* **206**, 153.

3WGA C. S. Wright (1987). Refinement of the crystal structure of wheat germ agglutinin isolectin 2 at 1.8 Å resolution. *J. Mol. Biol.* **194**, 501.

1WRP, 2WRP, 3WRP C. L. Lawson, R.-G. Zhang, R. W. Schevitz, Z. Otwinowski, A. Joachimiak, and P. B. Sigler (1988). Flexibility of the DNA-binding domains of *trp* repressor. *Proteins: Struct., Funct., Genet.* **3**, 18.

1WSY C. C. Hyde, S. A. Ahmed, E. A. Padlan, E. W. Miles, and D. R. Davies (1988). Three-dimensional structure of the tryptophan synthase $\alpha_2\beta_2$ multienzyme complex from *Salmonella typhimurium*. *J. Biol. Chem.* **263**, 17857.

1XIA, 2XIA K. Henrick, D. M. Blow, H. L. Carrell, and J. P. Glusker (1987). Comparison of backbone structures of glucose isomerase from *Streptomyces* and *Arthrobacter*. *Protein Eng.* **1**, 467.

3XIA G. K. Farber, A. Glasfeld, G. Tiraby, D. Ringe, and G. A. Petsko (1989). Crystallographic studies of the mechanism of xylose isomerase. *Biochemistry* **28**, 7289.

4XIA, 5XIA K. Henrick, C. A. Collyer, and D. M. Blow (1989). Structures of D-xylose isomerase from *Arthrobacter* strain B3728 containing the inhibitors xylitol and D-sorbitol at 2.4 Å and 2.3 Å resolution, respectively. *J. Mol. Biol.* **208**, 129.

1XY1, 1XY2 S. P. Wood, I. J. Tickle, A. M. Treharne, J. E. Pitts, Y. Mascarenhas, J. Y. Li, J. Husain, S. Cooper, T. L. Blundell, V. J. Hruby, A. Buku, A. J. Fischman, and H. R. Wyssbrod (1986). Crystal structure analysis of deamino-oxytocin. Conformational flexibility and receptor binding. *Science* **232**, 633.

2YHX C. M. Anderson, R. E. Stenkamp, and T. A. Steitz (1978). Sequencing a protein by x-ray crystallography. II. Refinement of yeast hexokinase B. Co-ordinates and sequence at 2.1 Å resolution. *J. Mol. Biol.* **123**, 15.

1ZNA H. R. Drew and R. E. Dickerson (1981). Conformation and dynamics in a Z-DNA tetramer. *J. Mol. Biol.* **152**, 723.

2ZNA, 3ZNA A. H.-J. Wang, G. J. Quigley, F. J. Kolpak, G. can der Marel, J. H. van Boom, and A. Rich. Left-handed double helical DNA. Variations in the backbone conformation. *Science* **211**, 171.

Tables of entries in the Protein Data Bank

1. Classed by function

NON-ENZYMES (including viruses)

ANTI-HYPERTENSIVE, ANTI-VIRAL
  BDS-I   1BDS,2BDS

CALCIUM BINDING PROTEIN
  Calmodulin (Bovine)   2CLN
  Calmodulin (Rat)   3CLN
  alpha-Lactalbumin   1ALC

COAGULATION INHIBITOR
  Hirudin (Hirudo medicinalis)   2HIR,4HIR,5HIR,6HIR

CONTRACTILE SYSTEM PROTEIN
  Calcium-binding parvalbumin b   1CPV,2CPV,3CPV
  Calcium-binding protein (intestinal)   3ICB
  Tropomyosin (Rabbit)   2TMA
  Troponin C (Chicken)   4TNC
  Troponin C (Rabbit)   1TNC
  Troponin C (Turkey)   5TNC

CRYSTALLIN
  γ-Crystallin   1GCR

CYTOKINE
  Interleukin-1β   1I1B,2I1B,4I1B

DNA BINDING
  434 Repressor protein   1R69
  Catabolite gene activator protein   2GAP,3GAP
  Cro repressor protein (Phage lambda)   1CRO
  Cro repressor protein (Phage 434)   2CRO
  Gene 5 DNA unwinding protein   2GN5
  Lambda repressor (N-terminal domain)   1LRP,1LRD
  Trp repressor protein (Escherichia coli)   1WRP,2WRP,3WRP

ELECTRON TRANSFER (cuproprotein)
  Azurin (Alcaligenes denitrificans)   2AZA
  Azurin (Pseudomonas aeruginosa)   1AZU
  Plastocyanin   1PCY,2PCY,3PCY,4PCY,5PCY,6PCY
  Pseudoazurin (Alcaligenes faecalis)   1PAZ,2PAZ

ELECTRON TRANSFER (cytochrome)
  Cytochrome b5   2B5C
  Cytochrome b562   156B
  Cytochrome c (Albacore)   3CYT,5CYT
  Cytochrome c (Bonito)   1CYC
  Cytochrome c (Rice)   1CCR
  Cytochrome c'   2CCY
  Cytochrome c2   2C2C,3C2C
  Cytochrome c3 (Desulfovibrio desulfuricans Norway)   1CY3
  Cytochrome c3 (Desulfovibrio vulgaris)   2CDV
  Cytochrome c5 (Azotobacter vinelandii)   1CC5
  Cytochrome c550   155C
  Cytochrome c551   351C,451C

ELECTRON TRANSFER (flavoprotein)
    Flavodoxin (Clostridium MP)   3FXN,4FXN
    Flavodoxin (Desulfovibrio vulgaris)   1FX1

ELECTRON TRANSFER (iron-sulfur protein)
    Ferredoxin (Azotobacter vinelandii)   4FD1,1FD2
    Ferredoxin (Bacillus thermoproteolyticus)   1FXB
    Ferredoxin (Peptococcus aerogenes)   1FDX
    Ferredoxin (Spirulina platensis)   3FXC
    High potential iron protein (HIPIP)   1HIP
    Rubredoxin (Clostridium pasteurianum)   4RXN,5RXN
    Rubredoxin (Desulfovibrio gigas)   1RDG
    Rubredoxin (Desulfovibrio vulgaris)   3RXN

ELONGATION FACTOR
    Elongation factor - tu   1EFM,1ETU

EXCITATION ENERGY TRANSFER
    Bacteriochlorophyll-a protein   3BCL

GLYCOSIDASE INHIBITOR
    alpha-amylase inhibitor HOE-467A   1HOE
    Tendamistat   2AIT

HORMONE
    Avian pancreatic polypeptide   1PPT
    p-Mercaptopropionate oxytocin   1XY1,1XY2
    Glucagon   1GCN
    Insulin (Bovine)   2INS
    Insulin (Porcine)   3INS,4INS
    Insulin-like growth factor I (IGF I) (somatomedin)   1GF1
    Insulin-like growth factor II (IGF II) (somatomedin)   1GF2
    Relaxin   1RLX,2RLX,3RLX,4RLX

HISTOCOMPATIBILITY ANTIGEN
    Histocompatibility antigen A2   1HLA,3HLA
    Histocompatibility antigen AW 68.1   2HLA

IMMUNOGLOBULIN
    Bence-Jones protein (Mcg)   1MCG
    Bence-Jones protein (Rei, variable fragment)   1REI
    Bence-Jones protein (Rhe, variable fragment)   2RHE
    IgG Fab fragment (J539)(galactan-binding)   1FBJ
    IgA Fab fragment (MCPC603)   1MCP,2MCP
    IgG1 Fab fragment (Kol)   2FB4
    IgG Fab' fragment (New)   3FAB
    IgG2 Fab fragment (R19.9)   1F19
    IgE Fc fragment   1IGE
    IgG1 Fc fragment   1FC1
    IgG1 Fc fragment and fragment b of protein A complex   1FC2
    IgG1 pFc' fragment   1PFC
    IgA Fv fragment (19.1.2)   1FVB,2FVB
    IgA Fv fragment (HyHEL-10)   1HFM
    IgA fv fragment (W3129)   1FVW,2FVW
    IgG1 (Kol)   2IG2

COMPLEX (immoglobulin-antigen)
    IgG1 Fab fragment (HyHEL-5) - lysozyme   2HFL
    IgA Fab fragment (HyHEL-10) - Lysozyme   3HFM
    IgA Fv fragment (HyHEL-10) - Lysozyme   2HFM

LECTIN
    Concanavalin A   2CNA,3CNA,1CN1
    Wheat germ agglutinin (isolectin 2)   3WGA

LIPID-ASSOCIATED PROTEIN
    Murein lipoprotein   1MLP

LYMPHOKINE
    Tumor Necrosis Factor-α (Human)   1TNF

MEMBRANE GLYCOPROTEIN
    Haemagglutinin   1HMG

NUCLEIC ACID ASSOCIATED PROTEIN
    Ubiquitin (Human)   1UBQ

ONCOGENE PROTEIN
    c-H-ras P21   2P21,3P21

OXYGEN STORAGE
  Myoglobin (Aplysia limacina)   1MBA,2MBA,3MBA,4MBA
  Myoglobin (Porcine)   1PMB
  Myoglobin (Seal)   1MBS
  Myoglobin (Sperm whale)   1MBD,1MBN,4MBN,5MBN,
                                             1MBO,1MB5,1MBC

OXYGEN TRANSPORT
  Hemerythrin b (Phascolopsis gouldii)   1HRB
  Hemerythrin (Siphonosoma funafati)   1HR3
  Hemerythrin (Themiste dyscritum)   1HMQ,1HMZ
  Hemoblogin (Deer)   1HDS
  Hemoglobin (erythrocruorin)   1ECA,1ECD,1ECN,1ECO
  Hemoglobin (Horse)   2DHB,2MHB
  Hemoglobin (Human)   1COH,1HCO,2HCO,2HHB,3HHB,
                                             4HHB,1HHO
  Hemoglobin (Human, fetal F II)   1FDH
  Hemoglobin (Lamprey)   2LHB
  Hemoglobin s (Human)   1HBS
  Leghemoglobin   1LH1,2LH1,1LH2,2LH2,1LH3,2LH3,
                                             1LH4,2LH4,1LH5,2LH5,1LH6,2LH6,
                                             1LH7,2LH7
  Myohemerythrin (Themiste zostericola)   2MHR

PEPTIDE ANTIBIOTIC
  Alamethicin   1AMT

PERIPLASMIC BINDING PROTEIN
  L-Arabinose-binding protein   1ABP
  D-Galactose-binding protein (Salmonella typhimurium)   1GBP
  Leucine-binding protein (Escherichia coli)   2LBP
  Leu-Ile-Val-binding protein (Escherichia coli)   2LIV

PHOTORECEPTOR
  Photoactive yellow protein (Ectothiorhodospira halophila)   1PHY

PHOTOSYNTHETIC REACTION CENTER
  Photosynthetic reaction center (Rhodopseudomonas viridis)   1PRC

PLANT SEED PROTEIN
  Crambin   1CRN

PROTEINASE INHIBITOR
  Chymotrypsin inhibitor CI-2   2CI2
  α_1 Proteinase inhibitor   5API,6API
  Ovomucoid third domain (Japanese quail)   1OVO
  Ovomucoid third domain (Silver pheasant)   2OVO
  Streptomyces subtilisin inhibitor   2SSI
  Trypsin inhibitor   4PTI,5PTI,6PTI

RIBOSOMAL PROTEIN
  L7/L12 50s ribosomal protein (C-terminal domain)   1CTF

STEROID BINING PROTEIN
  Uteroglobin   1UTG,2UTG

SWEET TASTING PROTEIN
  Monellin   1MON
  Thaumatin I   1THI

TOXIN
  Actinoxanthin   1ACX
  α-Bungarotoxin   2ABX
  α-Cobratoxin   1CTX
  Erabutoxin A (Sea snake)   5EBX
  Erabutoxin B (Sea snake)   3EBX
  Melittin   1MLT
  Neurotoxin b (probably identical to erabutoxin b)   1NXB
  Scorpion neurotoxin (variant 3)   1SN3

TRANSPORT (thyroxine, retinol)
  Prealbumin (Human plasma)   2PAB

TYPE 1 COPPER PROTEIN
  Cucumber basic protein   1CBP

VIRUS
  Mengo virus   2MEV
  Poliovirus   2PLV

Rhinovirus 14 (Human)	1R08,2R04,2R06,2R07,4RHV,2RM2,1RMU,
	2RMU,2RR1,2RS1,2RS3,2RS5
Satellite tobacco necrosis virus	2STV
Southern bean mosaic virus coat protein	4SBV
Tomato bushy stunt virus	2TBV
Tobacco Mosaic Virus	2TMV

ENZYMES

HYDROLASE (acid proteinase)

Acid proteinase (Endothia parasitica)	4APE
Acid proteinase (Penicillium janthinellum)	2APP
Acid proteinase (Rhizopus chinensis)	2APR
Chymosin B (Bovine)	1CMS
HIV-1 Protease	1HVP,2HVP,3HVP,4HVP
Rous sarcoma virus protease	2RSP
Pepsin (Porcine)	2PEP,3PEP,4PEP
Pepsinogen (Porcine)	1PSG

COMPLEX (acid proteinase-inhibitor)

Acid protease (Rhizopus chinensis) - reduced peptide inhibitor	3APR

HYDROLASE (acting on acid anhydride)

Inorganic pyrophosphatase	1PYP

HYDROLASE (acting on C-N bonds other than peptide bonds)

β-Lactamase (Staphylococcus aureus)	1BLM

HYDROLASE (carboxylic esterase)

Phospholipase a_2 (Bovine)	1BP2,3BP2
Phospholipase a_2 (Porcine)	1P2P,3P2P
Phospholipase a_2 (Rattlesnake)	1PP2
Prophospholipase a_2 (Bovine)	2BP2

HYDROLASE (endoribonuclease)

Ribonuclease T$_1$	1RNT,2RNT,3RNT

HYDROLASE (O-glycosyl)

Cellobiohydrolase I (Trichoderma reesei)	1CBH,2CBH
Lysozyme (Bacteriophage T4)	2LZM,3LZM,1LYD
Lysozyme (Bacteriophage T4) mutants	1L01,1L02,1L03,1L04,1L05,1L06,
	1L07,1L08,1L09,1L10,1L11,1L12,
	1L13,1L14,1L15,1L16,1L17,1L18,
	1L19,1L20,1L21,1L22,1L23,1L24,
	1L25,1L26,1L27,1L28,1L29,1L30,
	1L31,1L32,1L33,1L34,1L35
Lysozyme (Hen egg white)	1LYM,1LYZ,2LYZ,3LYZ,4LYZ,5LYZ,
	6LYZ,7LYZ,8LYZ,1LZH,2LZH,1LZT,
	2LZT,2LYM,3LYM
Lysozyme (Human)	1LZ1
Lysozyme (Nam-Nag-Nam substrate only)	9LYZ
Lysozyme (Turkey)	1LZ2,2LZ2
Taka-amylase A	2TAA

HYDROLASE (metallo-carboxypeptidase)

Carboxypeptidase Aα	3CPA,5CPA
Carboxypeptidase Aα-potato carboxypeptidase A inhibitor complex	4CPA
Carboxypeptidase B	1CPB

HYDROLASE (metalloproteinase)

Thermolysin	3TLN,4TLN,5TLN,7TLN,1TLP,
	1TMN,2TMN,3TMN,4TMN,5TMN,6TMN,
	7TMN

HYDROLASE (phosphoric diester)

Staphylococcal nuclease	2SNS
Ribonuclease A	1RN3,5RSA,6RSA,7RSA,1RSM
Ribonuclease B	1RBB
Ribonuclease S	1RNS

HYDROLASE (serine proteinase and zymogen)

α-Lytic proteinase	2ALP,1P01,1P02,1P03,1P04,1P05,
	1P06,1P07,1P08,1P09,1P10
α-Chymotrypsin A	2CHA,4CHA,5CHA
γ-Chymotrypsin A	2GCH
Chymotrypsinogen A	1CHG,2CGA

Tosyl-elastase (Sus scrofa)	1EST
Elastase (Procine)	3EST
Neutrophil Elastase (Human)	1HNE
Kallikrein A	2PKA
Proteinase A (Streptomyces griseus)	2SGA
Proteinase K (Tritirachium album limber)	2PRK
Rat mast cell proteinase II	3RP2
Subtilisin (Bacillis subtilis)	1SBC,1SBT,2SBT
Tonin (Rattus rattus)	1TON
Trypsin (Bovine)	3PTB,2PTN,3PTN,4PTP,1TPO,1TPP, 1TRM,2TRM,1TLD
Trypsin (Streptomyces Griseus)	1SGT
Trypsinogen	2TGA,1TGB,1TGC,1TGN, 1TGT,2TGT

COMPLEX (serine proteinase-inhibitor)

α-Chymotrypsin - Ovomucoid third domain	1CHO
α-Chymotrypsin - phenylethane boronic acid complex	6CHA
Elastase - anilide inhibitor	2EST
Kallikrein A - bovine pancreatic trypsin inhibitor	2KAI
Proteinase A (Streptomyces griseus) - chymostatin	1SGC
Proteinase B (Streptomyces griseus) - ovomucoid inhibitor	3SGB
Subtilisin carlsberg - Eglin-C (Hirudo medicinalis)	1CSE,2SEC
Subtilisin BPN'- streptomyces subtilisin inhibitor	1SIC
Subtilisin novo - Chymotrypsin inhibitor 2	2SNI
Thermitase - Eglin-C	1TEC
Trypsin - monoisopropylphosphoryl inhibited	1NTP
Trypsin - bovine pancreatic trypsin inhibitor	2PTC,1TPA
Trypsinogen - bovine pancreatic trypsin inhibitor	2TGP,2TPI,3TPI
Trypsinogen - DIP inhibited	2TGD
Trypsinogen - porcine pancreatic secretory trypsin inhibitor	1TGS
Trypsinogen - [Arg][15] pancreatic trypsin inhibitor - (Ile, Val) ternary complex	4TPI

HYDROLASE (sulfhydryl proteinase)

Actinidin	2ACT
Papain	1PAD,2PAD,4PAD, 5PAD,6PAD,9PAP
Papain D	1PPD

HYDROLASE (transpeptidase)

D-Alanyl-carboxypeptidase-transpeptidase	1PTE

ISOMERASE

D-Xylose isomerase (Arthrobacter)	1XIA,4XIA,5XIA
D-Xylose isomerase (Streptomyces olivochromogenes)	3XIA
D-Xylose isomerase (Streptomyces rubiginosus)	2XIA
D-Glucose-6-phosphate isomerase	1PGI
Triose phosphate isomerase (Chicken)	1TIM
D-Xylose isomerase	

LIGASE (amide synthetase)

Glutamine synthetase	2GLS

LIGASE (synthetase)

Tyrosyl-tRNA synthetase	2TS1,3TS1,4TS1

LYASE (oxo-acid)

Carbonic anhydrase form b (carbonate dehydratase)	2CAB
Carbonic anhydrase form c (carbonate dehydratase)	1CA2,2CA2,3CA2
Citrate synthase (Chicken)	3CTS
Citrate synthase (Porcine)	1CTS,2CTS,4CTS
Enolase	2ENL
2-Keto-3-deoxy-6-phosphogluconate (KDPG) aldolase	1KGA

LYASE (carbon-oxygen)

RuBisCo (Rhodospirillum rubrum)	2RUB
Tryptophan Synthase (Salmonella typhimurium)	1WSY

OXIDOREDUCTASE (acting on paired donors with incorporation of O_2)

p-Hydroxybenzoate hydroxylase	1PHH

OXIDOREDUCTASE (aldehyde(d)-NAD(a))

D-Glyceraldehyde-3-phosphate dehydrogenase (Bacillus stearothermophilus)	1GD1,2GD1
D-Glyceraldehyde-3-phosphate dehydrogenase (Human)	3GPD
D-Glyceraldehyde-3-phosphate dehydrogenase (Lobster)	1GPD,4GPD

207

OXIDOREDUCTASE (CH-NH(d)-NAD$^{(+)}$ or NADP$^{(+)}$(a))
Dihydrofolate reductase (Chicken)	8DFR
Dihydrofolate reductase (Escherichia coli)	4DFR
Dihydrofolate reductase (Lactobacillus casei)	3DFR

OXIDOREDUCTASE (NADPH(d)-disulfide(a))
Glutathione reductase	3GRS
Thioredoxin reductase	1SRX

OXIDOREDUCTASE (H$_2$O$_2$(a))
Catalase (Beef liver)	7CAT,8CAT
Catalase (Penicillium vitale)	4CAT
Cytochrome c peroxidase	2CYP
Glutathione peroxidase (Bovine)	1GP1

OXIDOREDUCTASE (CHOH(d)-NAD(a))
Cytoplasmic malate dehydrogenase	4MDH
Lactate dehydrogenase (Dogfish)	3LDH,6LDH,8LDH,1LDM
Lactate dehydrogenase (Mouse)	2LDX
Lactate dehydrogenase (Porcine)	5LDH
Lactate dehydrogenase (Bacillus stearothermophilus)	1LDB,2LDB
Lactate dehydrogenase (Lactobacillus casei)	1LLC
Liver alcohol dehydrogenase	5ADH,6ADH,7ADH,8ADH

OXIDOREDUCTASE (Oxygen(a))
Glycolate Oxidase	1GOX

OXIDOREDUCTASE (oxygenase)
Cytochrome P450CAM (Pseudomonas putida)	2CPP,3CPP

OXIDOREDUCTASE (superoxide radical(a))
Cu,Zn superoxide dismutase	2SOD

TRANSFERASE (aminotransferase)
Cytosolic aspartate aminotransferase (Chicken)	1AAT
Cytosolic aspartate aminotransferase (Escherichia coli)	2AAT

TRANSFERASE (carbamoyl transferase)
Aspartate carbamoyltransferase	2ATC,4ATC,7ATC

TRANSFERASE (nucleotidyltransferase)
DNA polymerase I	1DPI

TRANSFERASE (phosphotransferase)
Adenylate kinase	3ADK
cAMP-dependent Protein Kinase (Bovine)	1APK,2APK,1BPK,2BPK
Hexokinase a	1HKG
Hexokinase b	2YHX
Phosphofructokinase (Bacillus stearothermophilus)	3PFK,4PFK,5PFK
Phosphofructokinase (Escherichia coli)	1PFK,2PFK
Phosphoglycerate kinase (Horse)	2PGK
Phosphoglycerate kinase (Yeast)	3PGK
Phosphoglycerate mutase	3PGM
Pyruvate kinase	1PYK

TRANSFERASE (sulfur transferase)
Rhodanese	1RHD

RNA

TRANSFER RIBONUCLEIC ACID
tRNA Asp (Yeast)	2TRA,3TRA
tRNA Phe (Yeast)	4TNA,6TNA,1TRA,1TN1,1TN2,4TRA

DNA

DEOXYRIBONUCLEIC ACID
DNA(ATCGGCTAAG) (model for DNA in chromatin)	1DNN
DNA (CCAAGATTGC)	3DNB
DNA (CCAGGCCTGG)	1BD1
A-DNA (CCGG)	1ANA
DNA (CGCAAATTTGCG)	2DND
DNA (CGCATATATGCG)	1DN9

B-DNA (CGCGAATTCGCG)	1BNA,2BNA,3BNA,4BNA, 5BNA,6BNA,7BNA,8BNA,1DNH
DNA (CGCG+AATTCGCG)	4DNB
DNA (CGCGATATCGCG)	1DNE
DNA (CGCGCG)	1DCG,2DCG
DNA (+CG+CG+CG)	1DN4,1DN5
DNA (CGCGCGTTTTCGCGCG)	1D16
DNA (CGTACGTACG) - Co(NH3) 6^{+++}	1DN8
DNA (GCCCGGGC)	9DNA
DNA (GGATGGGAG)	1DN6
DNA (GGGATCCC)	3ANA
DNA (GTGTACAC)	1DNS
DNA (GTACGTAC)	5ANA
A-DNA (GGGGCCCC)	2ANA
Poly (dGEdC)	1DN7
Z-DNA (CGCG)	1ZNA
Z-I-DNA (CGCGCGCGCGCG)	2ZNA
Z-II-DNA (CGCGCGCGCGCG)	3ZNA

POLYSACCHARIDES

POLYSACCHARIDES

Agarose	1AGA
Capsular polysaccharide	1CAP
Chondroitin sulfate	1C4S,2C4S
Hyaluronic acid	1HYA,2HYA,3HYA,4HYA
Iota-Carrageenan	1CAR
Keratan sulfate	1KES

NOTES:

1. This table is current through April, 1990.
2. More than one entry for a given macromolecule may occur in the Data Bank. These multiple entries pertain to:

   a) Work done on the same macromolecule in different laboratories.
   b) Structure analysis using different refinement programs.
   c) Complexes with small molecules.
   d) Different oxidation states of bound ligands (e.g. reduced/oxidized cytochromes).
   e) Chemical modification of side chains.
   f) Mutants.

2. Sorted in order of Protein Data Bank code, ignoring the initial numeral; bibliographic entries first

This information has been extracted directly from the entries, without deletion of special characters indicating capitalization, etc. (e.g., /DNA*). The reason for this is to facilitate searching for text in the original data base or in other data bases derived from it. N.A. in the column under Resolution stands for 'Not Applicable', as for a hypothetical model of a structure.

PDB CODE		COMPOUND	SOURCE	RESOLUTION
1	0SC1	CYTOCHROME $C-555=	(CHLOROBIUM THIOSULFATOPHILUM)CHLOROBIUM LIMICOLA F.SP.	1.5
2	0ACD	ACYL-*CO*A DEHYDROGENASE (E.C.1.3.99.3)	PIG (SUS $SCROFA $DOMESTICA) LIVER MITOCHONDRIA	3.0
3	0ACI	ASPARTATE CARBAMOYLTRANSFERASE (/ACT$ASE)(E.C.2.1.3.2) (I CRYSTAL)	(ESCHERICHIA $COLI) STRAIN E*K1104 CONTAINING THE PLASM	3.0
4	0ACS	ASPARTATE CARBAMOYLTRANSFERASE (/ACT$ASE)(E.C.2.1.3.2) -*CARBAMOYL PHO	(ESCHERICHIA $COLI)	2.6
5	0AD2	ADENOVIRUS TYPE 2 HEXON ($AD2)	HUMAN (HOMO $SAPIENS)	2.9
6	0AF1	APOFERRITIN	HORSE (EQUUS CABALLUS) SPLEEN	2.8
7	0AFP	ANTIFREEZE POLYPEPTIDE (A*F*P) (H*P*L*C-6)	WINTER FLOUNDER (PSEUDOPLEURONECTES $AMERICANUS)	2.5
8	0AKA	ADENYLATE KINASE (E.C.2.7.4.3) -*P=-1==,*P==5===-DI ($ADENOSINE-5 ($PRIME	(ESCHERICHIA $COLI)	2.3
9	0AKN	ADENYLATE KINASE (E.C.2.7.4.3)	YEAST (SACCHAROMYCES $CEREVISIAE) CYTOSOL	2.6
10	0ALD	ALDOLASE A (E.C.4.1.2.13)	RABBIT (ORYCTOLAGUS $CUNICULUS) SKELETAL MUSCLE	2.7
11	0AN8	/DNA$ A-(5*-D(GP-GP-TP-AP-TP-AP-CP-C)-3*)	SYNTHETIC /DNA$	2.25
12	0ANB	/DNA$ A-(5*-D(GP-GP-+*UP-AP-+*UP-AP-CP-C)-3*)	SYNTHETIC /DNA$	2.25
13	0AUI	IMMUNOGLOBULIN BENCE-*JONES (KAPPA)	HUMAN (HOMO $SAPIENS) FROM THE URINE OF THE MULTIPLE MYE	2.2
14	0B2M	BETA-2=MICROGLOBULIN	BOVINE (BOS $TAURUS) MILK AND COLOSTRUM	2.9
15	0BGT	ALPHA-*BUNGAROTOXIN	MANY BANDED KRAIT (BUNGARUS $MULTICINCTUS)	2.5
16	0C3A	SDES-*ARG==77=--C3A ANAPHYLATOXIN	HUMAN (HOMO $SAPIENS) SERUM	3.2
17	0CCI	CYTOCHROME $C PEROXIDASE COMPOUND I (E.C.1.11.1.5)	BAKER,$S YEAST (SACCHAROMYCES $CEREVISIAE)	2.1
18	0CDF	DIHYDROFOLATE REDUCTASE (E.C.1.5.1.3)	CHICKEN (GALLUS GALLUS) LIVER	2.90
19	0CDI	CALOTROPIN /DIS	MADAR PLANT (CALOTROPIS $GIGANTEA)	3.2
20	0CDT	CARDIOTOXIN V==*I*I=4	(NAJA $MOSSAMBICA $MOSSAMBICA)	3.0
21	0CHY	CHE*Y	(SALMONELLA $TYPHIMURIUM)	2.7
22	0CNA	CONCANAVALIN A (DEMETALLIZED)	JACK BEAN (CANAVALIA ENSIFORMIS)	2.8
23	0COL	COLICIN A (C-*TERMINAL DOMAIN)	(ESCHERICHIA $COLI)	2.5
24	0CPC	C-*PHYCOCYANIN	(AGMENELLUM $QUADRUPLICATUM)	2.5
25	0CPF	CYTOCHROME P450 (/CAM$) (E.C.1.14.15.1) SUBSTRATE-FREE	(PSEUDOMONAS $PUTIDA)	2.2
26	0CPP	CATABOLITE GENE ACTIVATOR PROTEIN 91 (/CAP91$)	(ESCHERICHIA $COLI) N*C*R91	2.4
27	0CPS	CARBOXYPEPTIDASE A=ALPHA= (COX) (E.C.3.4.17.1) COMPLEX WITH GLYCYL-*L-	BOVINE (BOS $TAURUS) PANCREAS	1.6
28	0CPT	CALCIUM-BINDING PARVALBUMIN /III$=$F= COMPLEX WITH TERBIUM	TOADFISH (OPSANUS $TAU) SWIMBLADDER MUSCLE	2.3
29	0CSB	STREPTAVIDIN-BIOTIN COMPLEX	(STREPTOMYCES $AVIDINII)	2.6
30	0CSE	SUBTILISIN CARLSBERG COMPLEX WITH EGLIN-C	(BACILLUS $SUBTILIS) AND LEECH (HIRUDO $MEDICINALIS)	1.2
31	0CYS	CYSTATIN	CHICKEN (GALLUS $GALLUS) EGG WHITE	2.0
32	0DAC	/DNA$-5(PRIME)-$D(P*GP*GP*GP*TP*AP*CP*GP*C)-3(PRIME)) COMPLEX WITH TRI	SYNTHETIC /DNA$	2.50
33	0DCH	HEMOGLOBIN (COBALT,DEOXY)	HUMAN (HOMO $SAPIENS)	2.50
34	0DF5	R67 DIHYDROFOLATE REDUCTASE (E.C.1.5.1.3) (TRIMETHOPRIM-RESISTANT)	(ESCHERICHIA $COLI) R-PLASMID R67 PRODUCT	2.8
35	0DN1	/DNA$ (5(PRIME)-$D(*CP*GP*GP*TP*CP*GP*C)-3(PRIME))	SYNTHETIC /DNA$	2.1
36	0DN2	/DNA$ (5(PRIME)- $D(*CP*GP*CP*AP*AP*TP*TP*CP*GP*CP*GP)-3(PRIME))	SYNTHETIC /DNA$	2.5
37	0DN3	/DNA$ (5(PRIME)- $D(*CP*GP*CP*GP*AP*AP*TP*TP*AP*GP*CP*GP)-3(PRIME))	SYNTHETIC /DNA$	2.5
38	0DN6	/DNA$-5(PRIME)-$D(P*GP*GP*AP*TP*GP*AP*GP*G)-3(PRIME))	SYNTHETIC /DNA$	3.0
39	0DNI	DEOXYRIBONUCLEASE I (D*NASE I) (E.C.3.1.21.1)	BOVINE (BOS $TAURUS) PANCREAS	2.0
40	0DRF	DIHYDROFOLATE REDUCTASE (E.C.1.5.1.3)-*FOLATE COMPLEX	HUMAN (HOMO $SAPIENS) RECOMBINANT FORM FROM (ESCHERICHI	2.0
41	0EAP	ACID PROTEINASE (E.C.3.4.23.10) ENDOTHIAPEPSIN	CHESTNUT BLIGHT FUNGUS (ENDOTHIA PARASITICA)	2.45
42	0EPC	ELASTASE (E.C.3.4.21.11) -(THR-*PRO-N*VAL-*N*ME*LEU-*TYR-*T DEGREES KE	PIG (SUS $SCROFA $DOMESTICA)	1.8
43	0ESC	ELASTASE (E.C.3.4.21.11) COMPLEX WITH TWO MOLECULES OF ACETYL-*ALA-*PR	PORCINE (SUS $SCROFA) PANCREAS	1.65
44	0ESZ	ELASTASE-N-*CARBOBENZOXY-*L-*ALANYL-P-NITROPHENYL ESTER COMPLEX (E.C.	PORCINE (SUS SCROFA) PANCREAS	3.5
45	0ETU	ELONGATION FACTOR TU-*GUANOSINE DIPHOSPHATE COMPLEX	(ESCHERICHIA COLI B)	2.6
46	0EVC	ELASTASE (E.C.3.4.21.37) (/HLE$) -*ME*O-*SUC-*ALA-*ALA-*PRO- ((MPCMK$)	HUMAN (HOMO $SAPIENS) LEUKOCYTE	2.3
47	0EXA	EXOTOXIN *A	(PSEUDOMONAS $AERUGINOSA)	3.0
48	0FCB	FLAVOCYTOCHROME $B-2= (E.C.1.1.2.3)	YEAST (SACCHAROMYCES $CEREVISIAE)	3.0
49	0FDL	FAB (IG*G D1.3)-LYSOZYME (E.C.3.2.1.17) COMPLEX	MOUSE (MUS $MUSCULUS) AND HEN (GALLUS $GALLUS) EGG WHIT	2.8
50	0FEI	FERREDOXIN I	(AZOTOBACTER $VINELANDII)	2.6

	PDB CODE	COMPOUND	SOURCE	RESOLUTION
51	0FMT	INITIATOR T/RNA=MET== =$F=	(ESCHERICHIA COLI)	3.5
52	0FX2	FLAVODOXIN (REDUCED FORM)	(CLOSTRIDIUM MP)	1.8
53	0FX3	FLAVODOXIN (OXIDIZED)	(ANACYSTIS $NIDULANS)	2.0
54	0FXI	FERREDOXIN I	(APHANOTHECE $SACRUM)	2.5
55	0GBP	D-GALACTOSE-BINDING PROTEIN	(ESCHERICHIA $COLI)	3.20
56	0GCB	GAMMA CHYMOTRYPSIN A (E.C.3.4.21.1) INACTIVATED BY 3-BENZYL-6-CHLORO-2	BOVINE (BOS $TAURUS) PANCREAS	1.9
57	0GCI	GAMMA-CHYMOTRYPSIN - 5-BENZYL-6-$CHLORO-2-$PYRONE COMPLEX	BOVINE (BOS $TAURUS) PANCREAS	1.50
58	0GD2	APO-+D-*GLYCERALDEHYDE-3-PHOSPHATE DEHYDROGENASE (E.C.1.2.1.12)	(BACILLUS $STEAROTHERMOPHILUS NCA 1503)	2.5
59	0GLM	LYSOZYME (E.C.3.2.1.17)	EMBDEN GOOSE (ANSER $ANSER)	2.10
60	0GLS	GLUTAMINE SYNTHETASE (E.C.6.3.1.2)	(SALMONELLA $TYPHIMURIUM)	3.5
61	0GOX	GLYCOLATE OXIDASE (E.C.1.1.3.1)	SPINACH (SPINACIA $OLERACIA)	2.2
62	0GTC	/DNA$ (A-5 (PRIME)-(G-G-G-C-T-C-C)-3(PRIME))	SYNTHETIC /DNA$	2.25
63	0HBG	HEMOGLOBIN	(GLYCERA DIBRANCHIATA)	2.5
64	0HBT	HEMOGLOBIN (T STATE) IN 20 PER CENT POLYETHYLENE GLYCOL (/PEG$)	HUMAN (HOMO $SAPIENS)	2.1
65	0HG1	HAEMAGGLUTININ (MUTANT D1112*G)	INFLUENZA VIRUS MUTANT D1112*G	3.0
66	0HG2	HAEMAGGLUTININ (MUTANT G146*D)	INFLUENZA VIRUS (MUTANT G146*D)	3.0
67	0HG3	HAEMAGGLUTININ (MUTANT L226*Q)	INFLUENZA VIRUS (MUTANT L226*Q)	2.9
68	0HG4	HAEMAGGLUTININ (MUTANT L226*Q)-*SIALIC ACID COMPLEX	INFLUENZA VIRUS (MUTANT L226*Q)	2.9
69	0HOE	ALPHA-*AMYLASE INHIBITOR NOE-467*A	(STREPTOMYCES $TENDAE 4158)	3.20
70	0HPI	HEMOCYANIN	(PANULIRUS $INTERRUPTUS)	3.3
71	0HRS	HEAVY RIBOFLAVIN SYNTHASE (/HRS$) (E.C.2.5.1.9)	(BACILLUS $SUBTILIS) MUTANT H94	3.3
72	0IG1	IMMUNOGLOBULIN G1 (KAPPA)	HUMAN (HOMO $SAPIENS) PLASMA FROM MULTIPLE MYELOMA PATIE	6.0
73	0ILB	INTERLEUKIN-1*BETA (/IL$-1*BETA)	HUMAN (HOMO $SAPIENS) RECOMBINANT FORM EXPRESSED IN (ES	2.4
74	0ILT	INTERLEUKIN-2	HUMAN (HOMO $SAPIENS)	1.1
75	0IN1	INSULIN	PIG (SUS SCROFA)	1.1
76	0IN2	INSULIN	PORCINE (SUS SCROFA) PANCREAS	1.8
77	0IN3	DESPENTAPEPTIDE INSULIN	BEEF (BOS $TAURUS)	1.50
78	0IN4	INSULIN	HUMAN (HOMO $SAPIENS)	2.8
79	0LPC	LIPOVITELLIN-*PHOSVITIN COMPLEX	LAMPREY (ICHTHYOMYZON $UNICUSPIS) YOLK	2.8
80	0LRP	N-TERMINAL DOMAIN OF LAMBDA REPRESSOR	(LAMBDA)	3.20
81	0LTN	LECTIN	PEA (PISUM $SATIVUM)	3.0
82	0LZ5	LYSOZYME (NEUTRON STUDY)	HEN (GALLUS GALLUS) EGG WHITE	3.0
83	0LZ6	LYSOZYME (E.C.3.2.1.17)	(STREPTOMYCES ERYTHRAEUS)	1.4
84	0LZE	DEUTERATED ETHANOL (CD3D2OH) LYSOZYME (NEUTRON STUDY) (E.C.3.2.1.17)	HEN (GALLUS $GALLUS) EGG WHITE	2.9
85	0LZG	LYSOZYME $G (/SEL$G) (GOOSE-TYPE)	BLACK SWAN (CYGNUS $ATRATUS) EGG WHITE	2.8
86	0LZT	LYSOZYME (E.C.3.2.1.17) (HIGH-TEMPERATURE FORM)	HEN (GALLUS $GALLUS) EGG WHITE	2.00
87	0MAA	MITOCHONDRIAL ASPARTATE AMINOTRANSFERASE (E.C.2.6.1.1)	CHICKEN (GALLUS GALLUS) HEART	2.8
88	0MB3	MYOGLOBIN (MET)	SPERM WHALE (PHYSETER CATODON)	1.8
89	0MBA	MYOGLOBIN	(APLYSIA $LIMACINA)	3.6
90	0MBC	MYOGLOBIN (FE II, CARBONMONOXY, 260 DEGREES K)	SPERM WHALE (PHYSETER $CATODON)	2.0
91	0MBM	MYOGLOBIN (MET)	SPERM WHALE (PHYSETER CATODON)	1.5
92	0MLE	MUCONATE LACTONIZING ENZYME (MUCONATE CYCLOISOMERASE I*I) (/MLE$) (E.C.	(PSEUDOMONAS $PUTIDA)	3.0
93	0MMD	MITOCHONDRIAL MALATE DEHYDROGENASE (E.C.1.1.1.38)	PORCINE (SUS $SCROFA) HEART	3.0
94	0MTS	METHIONYL-T/RNA$ SYNTHETASE (E.C.6.1.1.10)	(ESCHERICHIA COLI)	2.5
95	0PAL	PIKE 4 (DOT) 10 PARVALBUMIN BETA	PIKE (ESOX $LUCIUS) MUSCLE	1.93
96	0PB1	PHOSPHORYLASE $B (E.C.2.4.1.1)	RABBIT (ORYCTOLAGUS CUNICULUS) MUSCLE	3.0
97	0PEC	PAPAIN-*E-64 COMPLEX	PAPAYA (*CARICA $PAPAYA), AND (*ASPERGILLUS $JAPONICUS)	2.4
98	0PF1	PROTHROMBIN FRAGMENT 1	BOVINE (BOS $TAURUS)	2.8
99	0PFB	PLATELET FACTOR 4	BOVINE (BOS $PRIMIGENIUS $TAURUS)	3.0
100	0PFK	PHOSPHOFRUCTOKINASE (E.C.2.7.1.11)	(BACILLUS STEAROTHERMOPHILUS)	2.4

PDB CODE		COMPOUND	SOURCE	RESOLUTION
101	0PGL	PHOSPHOGLUCOMUTASE (E.C.2.7.5.1)	RABBIT (ORYCTOLAGUS $CUNICULUS) MUSCLE	2.7
102	0PHH	/P$-*HYDROXYBENZOATE HYDROXYLASE SUBSTRATE COMPLEX (E.C.1.14.13.2)	(PSEUDOMONAS FLUORESCENS)	2.5
103	0PLV	POLIOVIRUS	HUMAN (HOMO $APIENS) (TYPE1, MAHONEY STRAIN) POLIOVIRU	2.9
104	0PPA	PHOSPORYLASE $A (E.C.2.4.1.1)	RABBIT (ORYCTOLAGUS CUNICULUS) MUSCLE	2.5
105	0PRC	PHOTOSYNTHETIC REACTION CENTER	(RHODOPSEUDOMONAS $VIRIDIS)	3.0
106	0PSG	PEPSINOGEN	PORCINE (SUS $SCROFA)	1.8
107	0RBI	RIBONUCLEASE BI (BINASE)	(BACILLUS $INTERMEDIUS, STRAIN 7P)	3.20
108	0RBS	RIBONUCLEASE (/RN$ASE-/BS1$,E.C. NUMBER NOT ASSIGNED)	BOVINE (BOS $TAURUS) SEMINAL FLUID	2.5
109	0RCR	REACTION CENTER (R*C)	(RHODOBACTER $SPHAEROIDES) R-26	2.8
110	0REN	RENIN (E.C.3.4.23.15)	HUMAN (HOMO $APIENS) RECOMBINANT FORM	2.5
111	0RIA	RIBONUCLEASE A (E.C.3.1.4.22) COMPLEX WITH /DNA$ (5(PRIME)-$D(P*AP*AP*	BOVINE (BOS $TAURUS) PANCREAS	2.5
112	0RIC	RICIN (R*C*A=*I*I=)	CASTOR (RICINUS $COMMUNIS) SEED	2.8
113	0RIF	INTESTINAL FATTY ACID-BINDING PROTEIN	RAT (RATTUS $NORVEGICUS) EXPRESSED IN (ESCHERICHIA $COL	2.5
114	0RNB	BARNASE (E.C.3.4.21.15)	(BACILLUS AMYLOLIQUEFACIENS)	2.20
115	0ROY	IMMUNOGLOBULIN BENCE-*JONES (KAPPA) V-MONOMER	HUMAN (HOMO $APIENS) URINE MULTIPLE MYELOMA PATIENT /RO	3.0
116	0RPL	RIBOSOMAL PROTEIN L30	BOVINE (BOS $TAURUS) PANCREAS	2.0
117	0RSA	RIBONUCLEASE A (E.C.3.1.4.22) RIBONUCLEASE (EC 3.1.27.5), PANCREATIC	(STREPTOMYCES ERYTHREUS)	2.50
118	0RST	RIBONUCLEASE ST (E.C.2.7.7.26)	PORCINE (SUS $CROFA) CORPUS LUTEUM	2.50
119	0RX5	RELAXIN	(SALMONELLA $TYPHIMURIUM)	N.A.
120	0SBP	SULPHATE-BINDING PROTEIN	(ESCHERICHIA $COLI)	2.00
121	0SDE	FE-SUPEROXIDE DISMUTASE (E.C.1.15.1.1)	(THERMUS $THERMOPHILUS /HB$8)	3.10
122	0SDM	MANGANESE SUPEROXIDE DISMUTASE (E.C.1.15.1.1)	(PSEUDOMONAS $OVALIS)	2.4
123	0SDP	FE-SUPEROXIDE DISMUTASE (E.C.1.15.1.1)	(PSEUDOMONAS $OVALIS)	2.90
124	0SEC	SUBTILISIN CARLSBERG COMPLEX WITH EGLIN-C	(BACILLUS $SUBTILIS) AND LEECH (HIRUDO $MEDICINALIS)	2.0
125	0SNI	SUBTILISIN NOVO (E.C.3.4.21.14) COMPLEX WITH CHYMOTRYPSIN INHIBITOR 2	(BACILLUS $AMYLOLIQUEFACIENS) AND BARLEY (HORDEUM $VULG	2.1
126	0ST1	SUBTILISIN (B*A*S)	(BACILLUS $AMYLOLIQUEFACIENS)	1.8
127	0ST2	SUBTILISIN (/BAS$OX) (PEROXIDE-OXIDIZED)	(BACILLUS $AMYLOLIQUEFACIENS)	2.0
128	0TA1	TRANSFER RIBO-NUCLEIC ACID (YEAST,ASP) $T/RNA	YEAST (SACCHAROMYCES CEREVISIAE)	3.5
129	0TEC	THERMITASE-*EGLIN C COMPLEX	(THERMOACTINOMYCES $VULGARIS) AND LEECH (HIRUDO $MEDICI	2.5
130	0TEL	LYSOZYME (E.C.3.2.1.17)	TORTOISE (TRIONYX $GANGETICUS) EGG WHITE	1.60
131	0TFD	TRANSFERRIN (DIFERRIC)	RABBIT (ORYCTOLAGUS $CUNICULUS) PLASMA	3.3
132	0THI	THAUMATIN I	(THAUMATOCOCCUS $DANIELLII BENTH)	3.10
133	0TLL	THERMOLYSIN (E.C.3.4.24.4) COMPLEX WITH P-*LEU-/NH$2= (N-PHOSPHORYL-*	(BACILLUS $THERMOPROTEOLYTICUS)	1.6
134	0TLP	THERMOLYSIN (E.C.3.4.24.4) COMPLEX WITH PHOSPHORAMIDON (N-(*L-LEUCYL-	(BACILLUS $THERMOPROTEOLYTICUS)	2.3
135	0TMD	TRIMETHYLAMINE DEHYDROGENASE (E.C.1.5.99.7)	METHYLOTROPHIC BACTERIUM W=3=*A*=1=	2.4
136	0TMT	THERMITASE (E.C.3.4.21.14)	(THERMOACTINOMYCES $VULGARIS)	2.2
137	0TMV	TOBACCO MOSAIC VIRUS PROTEIN DISK	TOBACCO MOSAIC VIRUS	2.8
138	0TR1	TRANSFER RIBO-NUCLEIC ACID (YEAST, PHE), $T/RNA	YEAST (SACCHAROMYCES CEREVISIAE)	2.5
139	0TRO	$TRP REPRESSOR-O$PERATOR COMPLEX	(ESCHERICHIA $COLI) STRAIN W3110 $TRP L75 LEU CARRYING	2.4
140	0TT4	THIOREDOXIN REDUCTASE	BACTERIOPHAGE T4	2.8
141	0TTI	BETA TRYPSIN-*TRYPSIN INHIBITOR I (/CMTI-I$)	BOVINE (BOS $PRIMIGENIUS $TAURUS) $AND SQUASH (CUCURBIT	1.6
142	0UTG	UTEROGLOBIN	RABBIT (ORYCTOLAGUS CUNICULUS) FEMALE GENITAL TRACT	2.2
143	0WGI	WHEAT GERM AGGLUTININ (ISOLECTIN 1)	WHEAT (TRITICUM $VULGARIS) GERM	2.6
144	0WRP	$TRP REPRESSOR	(ESCHERICHIA $COLI)	3.0
145	0YPI	TRIOSE PHOSPHATE ISOMERASE	YEAST (SACCHAROMYCES $CEREVISIAE)	2.50
146	02GP	D-*ALANYL-*D-*ALANINE PEPTIDASE (ZN=2+== G PEPTIDASE) (E.C.3.4.17.8)	(STREPTOMYCES $ALBUS G)	1.8
147	02IN	INSULIN (2ZN-*INSULIN $PHENOL)	PIG (SUS $SCROFA $DOMESTICA)	1.6
148	351C	CYTOCHROME $C=551= (OXIDIZED)	(PSEUDOMONAS AERUGINOSA)	1.6
149	451C	CYTOCHROME $C=551= (REDUCED)	(PSEUDOMONAS AERUGINOSA)	1.6
150	155C	CYTOCHROME C550	(PARACOCCUS DENITRIFICANS) /ATCC 13543	2.5

PDB	CODE	COMPOUND	SOURCE	RESOLUTION
151	156B	CYTOCHROME B562 (E. COLI, OXIDIZED)CYTOCHROME B562 (E. COLI, OXIDIZED	(ESCHERICHIA COLI)	2.5
152	1AAT	CYTOSOLIC ASPARTATE AMINOTRANSFERASE (E.C.2.6.1.1) COMPLEX WITH 2-OXO-	CHICKEN (GALLUS GALLUS) HEART	2.8
153	2AAT	ASPARTATE AMINOTRANSFERASE (E.C.2.6.1.1) MUTANT K258A COMPLEX WITH PYR	(ESCHERICHIA $COLI)	2.8
154	1ABP	L-*ARABINOSE-BINDING PROTEIN	(ESCHERICHIA COLI)	2.4
155	2ABX	ALPHA-*BUNGAROTOXIN	BRAIDED KRAIT (BUNGARUS $MULTICINCTUS) VENOM	2.5
156	2ACT	ACTINIDIN (SULFHYDRYL PROTEINASE) (E.C. NUMBER NOT ASSIGNEDACTINIDIN (	CHINESE GOOSEBERRY OR KIWIFRUIT (ACTINIDIA CHINENSIS)	1.7
157	1ACX	ACTINOXANTHIN	(ACTINOMYCES GLOBISPORUS, NUMBER 1131)	2.0
158	5ADH	APO-LIVER ALCOHOL DEHYDROGENASE (E.C.1.1.1.1) COMPLEX WITH ADP-RIBOSE	HORSE (EQUUS $CABALLUS) LIVER	2.9
159	6ADH	HOLO-LIVER ALCOHOL DEHYDROGENASE (E.C.1.1.1.1) COMPLEX WITH NAD AND DM	HORSE (EQUUS $CABALLUS) LIVER	2.9
160	7ADH	ISONICOTINIMIDYLATED LIVER ALCOHOL DEHYDROGENASE (E.C.1.1.1.1)	HORSE (EQUUS $CABALLUS) LIVER	3.2
161	8ADH	APO-LIVER ALCOHOL DEHYDROGENASE (E.C.1.1.99.8)	HORSE (EQUUS $CABALLUS) LIVER	2.4
162	3ADK	ADENYLATE KINASE (E.C.2.7.4.3)	PORCINE (SUS $SCROFA) MUSCLE	2.1
163	1AGA	AGAROSE (AN ALTERNATING COPOLYMER OF 3-LINKED BETA-D-GALACT 3,6-ANHYDR	RED SEAWEED (RHODOPHYCAE). SAMPLES OF VARIOUS COMPOSIT	3.0
164	2AIT	TENDAMISTAT	(STREPTOMYCES $TENDAE)	N.A.
165	1ALC	ALPHA-*LACTALBUMIN	BABOON (PAPIO $CYNOCEPHALUS) MILK	1.7
166	2ALP	ALPHA-LYTIC PROTEASE (E.C.3.4.21.12)	(LYSOBACTER $ENZYMOGENES)	1.7
167	1AMT	ALAMETHICIN	(TRICHODERMA $VIRIDE)	1.5
168	1ANA	/DNA$ (A, 5(PRIME)-$D(=I=*CP*CP*GP*G)-3(PRIME))	SYNTHETIC DNA	2.1
169	2ANA	/DNA$ (A,5(PRIME)-$D(*GP*GP*GP*GP*CP*CP*CP*C)-3(PRIME))	SYNTHETIC /DNA$	2.5
170	3ANA	/DNA$-5(PRIME)-$D(GP*GP*AP*TP*CP*CP*C)-3(PRIME) (A CONFORMATION)	SYNTHETIC /DNA$	2.5
171	5ANA	/DNA$ (5(PRIME)-$D(GP*TP*AP*TP*AP*CP)-3(PRIME))	SYNTHETIC	2.25
172	4APE	ACID PROTEINASE (E.C.3.4.23.10), ENDOTHIAPEPSIN	CHESTNUT BLIGHT FUNGUS (ENDOTHIA $PARASITICA)	2.1
173	5API	MODIFIED ALPHA-1=-*ANTITRYPSIN (MODIFIED ALPHA-1=-*PROTEINASE INHIBITO	HUMAN (HOMO $SAPIENS)	3.0
174	6API	MODIFIED ALPHA-1=-*ANTITRYPSIN (MODIFIED ALPHA-1=-*PROTEINASE INHIBITO	HUMAN (HOMO $SAPIENS)	3.0
175	1APK	C/AMP DEPENDENT PROTEIN KINASE (EC 2.7.1.37) TYPE I, DOMAIN *A (MODE	BOVINE (BOS $PRIMIGENIUS $TAURUS) MUSCLE	N.A.
176	2APK	C/AMP DEPENDENT PROTEIN KINASE (E.C.2.7.1.37) TYPE /II$, DOMAIN *A (	BOVINE (BOS $PRIMIGENIUS $TAURUS) CARDIAC MUSCLE	N.A.
177	2APP	ACID PROTEINASE (E.C.3.4.23.7), PENICILLOPEPSIN	FUNGUS (PENICILLIUM JANTHINELLUM)	1.8
178	2APR	ACID PROTEINASE (RHIZOPUSPEPSIN) (E.C.3.4.23.6)	BREAD MOLD (RHIZOPUS $CHINENSIS)	1.8
179	3APR	ACID PROTEINASE (RHIZOPUSPEPSIN) (E.C.3.4.23.6) COMPLEX WIT (==5==PSI=	BREAD MOLD (RHIZOPUS $CHINENSIS)	1.8
180	2ATC	ASPARTATE CARBAMOYLTRANSFERASE (ASPARTATE TRANSCARBAMYLASE) (E.C.2.1.3	(ESCHERICHIA COLI)	3.0
181	4ATC	ASPARTATE CARBAMOYLTRANSFERASE (ASPARTATE TRANSCARBAMYLASE) (E.C.2.1.3	(ESCHERICHIA $COLI)	2.6
182	7ATC	/CTP$-LIGANDED ASPARTATE CARBAMOYLTRANSFERASE (ASPARTATE TRANSCARBAMYL	(ALCALIGENES $DENITRIFICANS, STRAIN /NCTC$ 8582)	2.6
183	2AZA	AZURIN (OXIDIZED)	(ALCALIGENES $DENITRIFICANS, STRAIN /NCTC$ 8582)	1.8
184	1AZU	AZURIN	(PSEUDOMONAS AERUGINOSA)	2.7
185	2B5C	CYTOCHROME B5 (OXIDIZED)	BOVINE (BOS TAURUS) LIVER, SOLUBILIZED FROM MICROSOMES	2.0
186	3BCL	BACTERIOCHLOROPHYLL-A PROTEIN	(PROSTHECOCHLORIS $AESTUARII, STRAIN 2K)	1.9
187	1BD1	B-/DNA$-5(PRIME)-$D(CP*CP*AP*GP*GP*CP*CP*TP*GP*G)-3(PRIME)	SYNTHETIC /DNA$	1.6
188	1BDS	/BDS-I$ (/NMR$, MINIMIZED MEAN STRUCTURE)	SEA ANEMONE (ANEMONIA $SULCATA)	N.A.
189	2BDS	/BDS-I$ (NMR, 42 SIMULATED ANNEALING STRUCTURES)	SEA ANEMONE (ANEMONIA $SULCATA)	N.A.
190	1BLM	BETA-*LACTAMASE (E.C.3.5.2.6)	(STAPHYLOCOCCUS $AUREUS /PC1$)	2.5
191	1BNA	/DNA$ (B, 5(PRIME)-$D(*CP*GP*CP*GP*AP*AP*TP*TP*CP*GP*CP*G)- (290 DEGRE	SYNTHETIC /DNA$	1.9
192	3BNA	/DNA$ (B, 5(PRIME)-$D(*CP*GP*CP*GP*AP*AP*TP*TP*CP*GP*CP*G)- (16 DEGREE	SYNTHETIC DNA	2.7
193	3BNA	/DNA$ (B, 5(PRIME)-$D(*CP*GP*CP*AP*AP*TP*TP*BR*CP*GP*CP (60 PER CE	SYNTHETIC DNA	3.0
194	4BNA	/DNA$ (B, 5(PRIME)-$D(*GP*GP*CP*AP*AP*TP*TP*BR*CP*GP*CP (60 PER CE	SYNTHETIC DNA	2.3
195	5BNA	/DNA$ (B, 5--$D(*CP*GP*CP*GP*AP*AP*TP*TP*CP*GP*CP*G)-3*(COMPLEX WITH	SYNTHETIC DNA	2.6
196	6BNA	/DNA$ (B, 5(PRIME)-$D(*CP*GP*CP*GP*AP*AP*TP*TP-==R==*CP*G COMPLEX WI	SYNTHETIC DNA	2.21
197	7BNA	/DNA$ (B, 5(PRIME)-$D(*CP*GP*CP*GP*AP*AP*TP*TP*CP*GP*CP*G)- (290 DEGRE	SYNTHETIC /DNA$	1.9
198	8BNA	/DNA$ (B, 5(PRIME)-$D(*CP*GP*CP*GP*AP*AP*TP*TP*CP*GP*CP*G) COMPLEX WI	SYNTHETIC DNA	2.2
199	1BP2	PHOSPHOLIPASE A2= (E.C.3.1.1.4) (PHOSPHATIDE ACYL-HYDROLASE)	BOVINE (BOS TAURUS L.) PANCREAS	1.7
200	2BP2	PROPHOSPHOLIPASE A2=	BOVINE (BOS TAURUS L.) PANCREAS	3.0

214

PDB	CODE	COMPOUND	SOURCE	RESOLUTION
201	3BP2	PHOSPHOLIPASE A2= (E.C.3.1.1.4) (PHOSPHATIDE ACYL-HYDROLASE) - TRANSA	BOVINE (BOS TAURUS L.) PANCREAS	2.1
202	1BPK	C/AMP DEPENDENT PROTEIN KINASE (EC 2.7.1.37) TYPE I, DOMAIN B (MODEL	BOVINE (BOS $PRIMIGENIUS $TAURUS) MUSCLE	N.A.
203	2BPK	C/AMP DEPENDENT PROTEIN KINASE (E.C.2.7.1.37) TYPE /II$, DOMAIN B (M	BOVINE (BOS $PRIMIGENIUS $TAURUS) CARDIAC MUSCLE	N.A.
204	2C2C	CYTOCHROME $C=2= (OXIDIZED)	(RHODOSPIRILLUM $RUBRUM)	2.0
205	3C2C	CYTOCHROME $C=2= (REDUCED)	(RHODOSPIRILLUM $RUBRUM)	1.68
206	1C4S	CHONDROITIN-4-SULFATE (AN ALTERNATING COPOLYMER OF BETA-D-G 2-DEOXY-2-	BOVINE (BOS TAURUS) NASAL SEPTA	3.0
207	2C4S	CHONDROITIN-4-SULFATE (AN ALTERNATING COPOLYMER OF BETA-D-G SALT	SWARM RAT CHONDROSARCOMA, STRAIN BUF/N	3.0
208	1CA2	CARBONIC ANHYDRASE /II$ (CARBONATE DEHYDRATASE) (/HCA II$) (E.C.4.2.1.	HUMAN (HOMO $SAPIENS) ERYTHROCYTES	2.0
209	2CA2	CARBONIC ANHYDRASE /II$ (CARBONATE DEHYDRATASE) (/HCA II$) (E.C.4.2.1.	HUMAN (HOMO $SAPIENS) ERYTHROCYTES	1.9
210	3CA2	CARBONIC ANHYDRASE /II$ (CARBONATE DEHYDRATASE) (/HCA II$) 3-MERCURI-4	HUMAN (HOMO $SAPIENS) ERYTHROCYTES	2.0
211	2CAB	CARBONIC ANHYDRASE FORM B (CARBONATE DEHYDRATASE) (E.C.4.2.1.1)	HUMAN (HOMO $SAPIENS) ERYTHROCYTES	2.0
212	1CAP	CAPSULAR POLYSACCHARIDE	(ESCHERICHIA COLI) MUTANT STRAIN M41 OF SEROTYPE K29	3.0
213	1CAR	IOTA CARRAGEENAN (AN ALTERNATING COPOLYMER OF 3, 6-ANHYDRO-A BETA-D-GAL	RED SEAWEED (RHODOPHYCAE). HANDSORTED SAMPLES OF STRAI	3.0
214	4CAT	CATALASE (E.C.1.11.1.6)	PENICILLIUM VITALE	2.0
215	7CAT	CATALASE (E.C.1.11.1.6)	BEEF (BOS $TAURUS) LIVER	2.5
216	8CAT	CATALASE (E.C.1.11.1.6)	BEEF (BOS $TAURUS) LIVER	2.5
217	1CBH	C-TERMINAL DOMAIN OF CELLOBIOHYDROLASE I (/CT-CBH$ I) (E.C.3.2.1.91) (	CHEMICALLY SYNTHESIZED POLYPEPTIDE USING THE SEQUENCE O	N.A.
218	2CBH	C-TERMINAL DOMAIN OF CELLOBIOHYDROLASE I (/CT-CBH$ I) (E.C. STRUCTURES	SYNTHETIC POLYPEPTIDE USING THE SEQUENCE OF /CT-CBH$ I	N.A.
219	1CBP	CUCUMBER BASIC PROTEIN	CUCUMBER (CUCUMIS $SATIVUS) SEEDLINGS	2.5
220	1CC5	CYTOCHROME C=5= (OXIDIZED)	(AZOTOBACTER $VINELANDII)	2.5
221	1CCR	CYTOCHROME C	RICE EMBRYOS (ORYZA $SATIVA L)	1.5
222	2CCY	CYTOCHROME $C (PRIME)	(RHODOSPIRILLUM $MOLISCHIANUM)	1.67
223	2CDV	CYTOCHROME $C=3=	(DESULFOVIBRIO $VULGARIS MIYAZAKI IAM 12604)DESULFOVIBR	1.8
224	2CGA	CHYMOTRYPSINOGEN *A	BOVINE (BOS $TAURUS) PANCREAS	1.8
225	2CHA	ALPHA CHYMOTRYPSIN A (TOSYLATED) (E.C.3.4.21.1)	COW (BOS TAURUS)	2.0
226	4CHA	ALPHA-CHYMOTRYPSIN A (E.C.3.4.21.1)	COW (BOS $TAURUS)	1.68
227	5CHA	ALPHA CHYMOTRYPSIN A (E.C.3.4.21.1)	COW (BOS $TAURUS)	1.68
228	6CHA	ALPHA CHYMOTRYPSIN A (E.C.3.4.21.1) COMPLEX WITH PHENYLETHANE BORONIC	COW (BOS $TAURUS)	1.8
229	1CHG	CHYMOTRYPSINOGEN ACHYMOTRYPSINOGEN A (EC 3.4.21.1)	COW (BOS TAURUS)	2.5
230	1CHO	ALPHA-CHYMOTRYPSIN (E.C.3.4.21.1) COMPLEX WITH TURKEY OVOMUCOID THIRD	BOVINE (BOS $TAURUS) PANCREAS AND TURKEY (MELEAGRIS $GA	1.8
231	2CI2	CHYMOTRYPSIN INHIBITOR 2 (/CI$=2)	BARLEY (HORDEUM $VULGARE, HIPROLY STRAIN) SEEDS	2.0
232	2CLN	N==2115== TRIMETHYLCALMODULIN COMPLEX WITH TRIFLUOPERAZINE (MODEL)	BOVINE (BOS $TAURUS) BRAIN	N.A.
233	3CLN	CALMODULIN	RAT (RATTUS $RATTUS) TESTIS	2.2
234	1CMS	CHYMOSIN B (FORMERLY KNOWN AS RENNIN) (E.C.3.4.23.4)	BOVINE (BOS $TAURUS) EXPRESSED IN (ESCHERICHIA $COLI)	2.3
235	1CN1	CONCANAVALIN A (DEMETALLIZED)	JACK BEAN (CANAVALIA ENSIFORMIS)	3.2
236	2CNA	CONCANAVALIN A	JACK BEAN (CANAVALIA ENSIFORMIS)	2.0
237	3CNA	CONCANAVALIN A	JACK BEAN (CANAVALIA ENSIFORMIS)	2.4
238	1COH	ALPHA-FERROUS-CARBONMONOXY, BETA-COBALTOUS-DEOXY HEMOGLOBIN (T STATE)	HUMAN (HOMO $SAPIENS)	2.9
239	3CPA	CARBOXYPEPTIDASE A=ALPHA= (COX) (E.C.3.4.17.1) COMPLEX WITH GLYCYL-*L-	BOVINE (BOS TAURUS) PANCREAS	2.0
240	4CPA	CARBOXYPEPTIDASE A=ALPHA= (COX) (E.C.3.4.17.1) COMPLEX WITH POTATO CAR	BOVINE (BOS TAURUS) PANCREAS AND RUSSETT-BURBANK POTATO	2.5
241	5CPA	CARBOXYPEPTIDASE A=ALPHA= (COX) (E.C.3.4.17.1)	BOVINE (BOS TAURUS) PANCREAS	1.54
242	1CPB	CARBOXYPEPTIDASE B (E.C.3.4.12.3) FRACTION II	BOVINE (BOS TAURUS) PANCREAS	2.8
243	2CPP	CYTOCHROME P450CAM (CAMPHOR MONOOXYGENASE) (E.C.1.14.15.1) WITH BOUND	(PSEUDOMONAS $PUTIDA)	1.63
244	3CPP	CYTOCHROME P450CAM (CAMPHOR MONOOXYGENASE) (E.C.1.14.15.1) - REDUCED C	(PSEUDOMONAS $PUTIDA)	1.9
245	1CPV	CALCIUM-BINDING PARVALBUMIN B	CARP (CYPRINUS CARPIO)	1.85
246	2CPV	CALCIUM-BINDING PARVALBUMIN B	CARP (CYPRINUS CARPIO)	1.85
247	3CPV	CALCIUM-BINDING PARVALBUMIN B	CARP (CYPRINUS CARPIO)	1.85
248	1CRN	CRAMBIN	ABYSSINIAN CABBAGE (CRAMBE ABYSSINICA) SEED	1.5
249	1CRO	CRO REPRESSOR	BACTERIOPHAGE (LAMBDA)	2.2
250	2CRO	434 CRO PROTEIN	PHAGE 434	2.35

PDB CODE		COMPOUND	SOURCE	RESOLUTION
251	1CSE	SUBTILISIN CARLSBERG (E.C.3.4.21.14) (COMMERCIAL PRODUCT FR WITH EGLIN	(BACILLUS $SUBTILIS) AND LEECH (HIRUDO $MEDICINALIS)	1.2
252	1CTF	L7(SLASH)*L12 50 S RIBOSOMAL PROTEIN (C-TERMINAL DOMAIN)	(ESCHERICHIA $COLI, /MRE$ 600)	1.7
253	1CTS	CITRATE SYNTHASE (E.C.4.1.3.7) - CITRATE COMPLEX	PIG (SUS $SCROFA) HEART	2.7
254	2CTS	CITRATE SYNTHASE (E.C.4.1.3.7) - (CO*A, CITRATE) COMPLEX	PIG (SUS $SCROFA) HEART	2.0
255	3CTS	CITRATE SYNTHASE (E.C.4.1.3.7) - (CO*A, CITRATE) COMPLEX	CHICKEN (GALLUS $GALLUS) HEART MUSCLE	1.7
256	4CTS	CITRATE SYNTHASE (E.C.4.1.3.7) - OXALOACETATE COMPLEX	PIG (SUS $SCROFA) HEART	2.9
257	1CTX	ALPHA COBRATOXIN	COBRA (NAJA NAJA SIAMENSIS)	2.5
258	1CY3	CYTOCHROME $C=3=	(DESULFOVIBRIO $DESULFURICANS NORWAY)	2.5
259	1CYC	FERROCYTOCHROME $C	BONITO (KATSUWONUS $PELAMIS, LINNAEUS)	2.3
260	2CYP	CYTOCHROME $C PEROXIDASE (E.C.1.11.1.5) (FERROCYTOCHROME $C (COLON) H2	BAKER,S YEAST (SACCHAROMYCES $CEREVISIAE)	1.7
261	3CYT	CYTOCHROME $C (OXIDIZED)	ALBACORE TUNA (THUNNUS ALALUNGA) HEART	1.8
262	5CYT	CYTOCHROME $C (REDUCED)	ALBACORE TUNA (THUNNUS $ALALUNGA) HEART	1.5
263	1D16	/DNA$-5(PRIME) - $D(CP*GP*CP*GP*CP*GP*CP*GP*TP*TP*TP*TP*CP*GP*CP*GP*CP*GP*CP*G)-3(P	SYNTHETIC /DNA$	2.1
264	1DCG	/DNA$-5(PRIME) -$D(CP*GP*CP*GP*CP*G)-3(PRIME) COMPLEX WITH MAGNESIUM	SYNTHETIC /DNA$	1.0
265	2DCG	/DNA$-5(PRIME)-$D(CP*GP*CP*GP*CP*G)-3(PRIME) COMPLEX WITH MAGNESIUM AN	SYNTHETIC /DNA$	0.9
266	3DFR	DIHYDROFOLATE REDUCTASE (E.C.1.5.1.3) COMPLEX WITH NADPH AN METHOTREXA	(LACTOBACILLUS CASEI), DICHLOROMETHOTREXATE-RESISTANT S	1.7
267	4DFR	DIHYDROFOLATE REDUCTASE (E.C.1.5.1.3) COMPLEX WITH METHOTREXATE	(ESCHERICHIA COLI B), STRAIN /MB1428$, A METHOTREXATE-R	1.7
268	8DFR	DIHYDROFOLATE REDUCTASE (E.C.1.5.1.3)	CHICKEN (GALLUS $GALLUS) LIVER	1.7
269	2DHB	HEMOGLOBIN (HORSE,DEOXY)	HORSE (EQUUS CABALLUS)	2.8
270	1DN4	/DNA$-5(PRIME)-$D(5*BR*CP*GP5*BR*CP*GP*CP*GP*C)-3(PRIME)) (18 DEGREE	SYNTHETIC /DNA$	1.4
271	1DN5	/DNA$-5(PRIME)-$D(5*BR*CP*GP5*BR*CP*GP*CP*GP*C)-3(PRIME)) (37 DEGREE	SYNTHETIC /DNA$	1.4
272	1DN6	/DNA$-5(PRIME)-$D(GP*GP*AP*TP*AP*GP*GP*AP*G)-3(PRIME)	SYNTHETIC /DNA$	3.0
273	1DN7	/DNA$-5(PRIME)-$D(GP*GP*GP*GP*GP*GP*GP*GP*G -3(PRIME)	SYNTHETIC	N.A.
274	1DN8	/DNA$-5(PRIME)-$D($P*CP*GP*TP*AP*CP*G) - COBALT HEXAMMINE	SYNTHETIC /DNA$	1.5
275	1DN9	/DNA$ (5(PRIME)-$D(CP*GP*CP*AP*TP*AP*TP*AP*TP*GP*CP*GP) -3(PRIME)	SYNTHETIC	2.2
276	9DNA	A-/DNA$-5(PRIME)-$D(CP*CP*AP*AP*GP*AP*TP*TP*GP*G)-3(PRIME)	SYNTHETIC /DNA$	1.3
277	3DNB	/DNA$-5(PRIME)-$D(CP*GP*TP*AP*GP*AP*TP*TP*GP*G)-3(PRIME)	SYNTHETIC /DNA$	2.2
278	4DNB	/DNA$-5(PRIME)- $D(CP*GP*CP*GP*APM=6==*AP*TP*TP*CP*GP*CP*G)-3(PRIME)	SYNTHETIC /DNA$	2.2
279	2DND	/DNA$-5(PRIME)- $D(*CP*GP*CP*AP*AP*AP*AP*TP*TP*TP*CP*CP*G)-3P COMPLEX WI	SYNTHETIC /DNA$	2.4
280	1DNE	/DNA$-5(PRIME)- $D(*CP*GP*CP*AP*AP*AP*TP*TP*TP*CP*GP*CP*G)-3P COMPLEX WI	SYNTHETIC /DNA$	2.25
281	1DNH	/DNA$-5(PRIME)-$D(*CP*GP*CP*GP*AP*AP*TP*TP*CP*GP*CP*G)-3P COMPLEX WI	SYNTHETIC /DNA$	N.A.
282	1DNN	/DNA$ (5(PRIME)-$D((*AP*TP*CP*GP*CP*TP*AP*AP*G))- 3(PRIME)) MODEL	NOT APPLICABLE (MODEL STUDY)	2.0
283	1DNS	/DNA$ (5(PRIME)-$D(*GP*TP*GP*TP*AP*CP*AP*CP)) COMPLEX WITH SPERMINE	SYNTHETIC	2.8
284	1DPI	/DNA$ POLYMERASE I (KLENOW FRAGMENT) (E.C.2.7.7.7) - $D/CMP COMPLEX	(ESCHERICHIA $COLI)	2.8
285	3EBX	ERABUTOXIN $B	SEA SNAKE (LATICAUDA $SEMIFASCIATA) VENOM	1.4
286	5EBX	ERABUTOXIN $A	SEA SNAKE (LATICAUDA $SEMIFASCIATA) VENOM	2.0
287	1ECA	HEMOGLOBIN (ERYTHROCRUORIN, AQUO MET)	(CHIRONOMOUS THUMMI THUMMI) (FRACTION III)	1.4
288	1ECD	HEMOGLOBIN (ERYTHROCRUORIN, DEOXY)	(CHIRONOMOUS THUMMI THUMMI) (FRACTION III)	1.4
289	1ECN	HEMOGLOBIN (ERYTHROCRUORIN, CYANO MET)	(CHIRONOMOUS THUMMI THUMMI) (FRACTION III)	1.4
290	1ECO	HEMOGLOBIN (ERYTHROCRUORIN, CARBONMONOXY)	(CHIRONOMOUS THUMMI THUMMI) (FRACTION III)	1.4
291	1EFM	TRYPSIN-MODIFIED ELONGATION FACTOR TU (/EF$-*TU-/GDF$)	(ESCHERICHIA $COLI)	2.7
292	2ENL	ENOLASE (E.C.4.2.1.11) (2-PHOSPHO-*D-GLYCERATE HYDROLASE)	BAKER'S YEAST (SACCHAROMYCES $CEREVISIAE)	2.25
293	1EST	TOSYL-ELASTASE (E.C.3.4.21.11)	PORCINE (SUS SCROFA) PANCREAS	2.5
294	2EST	ELASTASE (E.C.3.4.21.11) COMPLEX WITH TRIFLUOROACETYL -*L-L (/TFAP$)	PORCINE (SUS $SCROFA) PANCREAS	2.5
295	3EST	NATIVE ELASTASE (E.C.3.4.21.11)	PORCINE (SUS $SCROFA) PANCREAS	1.65
296	1ETU	ELONGATION FACTOR TU (DOMAIN I) - *GUANOSINE DIPHOSPHATE COMPLEX	(ESCHERICHIA $COLI B)	2.9
297	1F19	R19.9 (IG*G2B=K=, /CRI$====-A=) FAB FRAGMENT	MOUSE (MUS $MUSCULUS) MONOCLONAL ANTIBODIES	2.8
298	3FAB	LAMBDA IMMUNOGLOBULIN FAB (PRIME)	HUMAN (HOMO $SAPIENS) PATIENT NEW	2.0
299	2FB4	IMMUNOGLOBULIN FAB	HUMAN (HOMO $SAPIENS) MYELOMA PATIENT KOL SERUM	1.9
300	1FBJ	IG*A FAB FRAGMENT (J539) (GALACTAN-BINDING)	MOUSE (MUS $MUSCULUS)	2.6

PDB	CODE	COMPOUND	SOURCE	RESOLUTION
301	1FC1	FC FRAGMENT (IGG1 CLASS)	HUMAN (HOMO SAPIENS) POOLED SERUM	2.9
302	1FC2	IMMUNOGLOBULIN FC AND FRAGMENT B OF PROTEIN A COMPLEX	HUMAN (HOMO SAPIENS) POOLED SERUM AND (STAPHYLOCOCCUS A	2.8
303	4FD1	FERREDOXIN	(AZOTOBACTER $VINELANDII /OP$) (/ATCC$ 13705)	1.9
304	1FD2	FERREDOXIN (MUTANT WITH CYS 20 REPLACED BY ALA) (/C20A$)	(AZOTOBACTER $VINELANDII /OP$) (/ATCC$ 13705)	1.9
305	1FDH	HEMOGLOBIN (DEOXY, HUMAN FETAL F=/II$=)	HUMAN FETUS (HOMO SAPIENS)	2.5
306	1FDX	FERREDOXIN	(PEPTOCOCCUS AEROGENES)	2.0
307	1FVB	IG*A FV FRAGMENT (19.1.2, ANTI-ALPHA(1(RIGHT ARROW)6) DEXTRAN) (MODEL)	MOUSE (MUS $MUSCULUS)	N.A.
308	2FVB	IG*A FV FRAGMENT (19.1.2, ANTI-ALPHA(1(RIGHT ARROW)6) DEXTRAN) (ENERGY	MOUSE (MUS $MUSCULUS)	N.A.
309	1FVW	IG*A FV FRAGMENT (W3129, ANTI-ALPHA(1(RIGHT ARROW)6) DEXTRAN) (MODEL)	MOUSE (MUS $MUSCULUS)	N.A.
310	2FVW	IG*A FV FRAGMENT (W3129, ANTI-ALPHA(1(RIGHT ARROW)6) DEXTRAN) (ENERGY-	MOUSE (MUS $MUSCULUS)	N.A.
311	1FX1	FLAVODOXIN	(DESULFOVIBRIO $VULGARIS)	2.0
312	1FXB	FERREDOXIN	(BACILLUS $THERMOPROTEOLYTICUS)	2.3
313	3FXC	FERREDOXIN	(SPIRULINA PLATENSIS)	2.5
314	3FXN	FLAVODOXIN (OXIDIZED FORM)	(CLOSTRIDIUM MP)	1.9
315	4FXN	FLAVODOXIN (SEMIQUINONE FORM)	(CLOSTRIDIUM MP)	1.8
316	2GAP	CATABOLITE GENE ACTIVATOR PROTEIN - DNA COMPLEX (MODEL)	(ESCHERICHIA $COLI)	N.A.
317	3GAP	CATABOLITE GENE ACTIVATOR PROTEIN - CYCLIC /AMP$ COMPLEX (/CAP$)	(ESCHERICHIA $COLI)	2.5
318	1GBP	GALACTOSE-BINDING PROTEIN	(SALMONELLA $TYPHIMURIUM STRAIN /ST1$)	3.0
319	2GCH	GAMMA CHYMOTRYPSIN A (E.C.3.4.21.1)	BOVINE (BOS TAURUS) PANCREAS	1.9
320	1GCN	GLUCAGON (PH 6 - PH 7 FORM)	PORCINE (SUS SCROFA) PANCREAS	3.0
321	1GCR	GAMMA-/II$ CRYSTALLIN	CALF (BOS $TAURUS) EYE LENS	1.6
322	1GD1	$HOLO-*D-*GLYCERALDEHYDE-3-PHOSPHATE DEHYDROGENASE (E.C.1.2.1.12)	(BACILLUS $STEAROTHERMOPHILUS /NCA$ 1503)	1.8
323	1GD1	$APO-*D-*GLYCERALDEHYDE-3-PHOSPHATE DEHYDROGENASE (E.C.1.2.1.12)	(BACILLUS $STEAROTHERMOPHILUS /NCA$ 1503)	2.5
324	1GF1	INSULIN-LIKE GROWTH FACTOR I (/IGF$ I) (SOMATOMEDIN)	HUMAN (HOMO SAPIENS)	N.A.
325	1GF2	INSULIN-LIKE GROWTH FACTOR /II$ (/IGF$ /II$) (SOMATOMEDIN)	HUMAN (HOMO SAPIENS)	N.A.
326	2GLS	GLUTAMINE SYNTHETASE (E.C.6.3.1.2)	(SALMONELLA $TYPHIMURIUM)	3.5
327	2GN5	GENE 5 /DNA$ BINDING PROTEIN	FILAMENTOUS BACTERIOPHAGE $FD (M13)	2.3
328	1GOX	GLYCOLATE OXIDASE (E.C.1.1.3.1)	SPINACH (SPINACIA $OLERACEA)	2.0
329	1GP1	GLUTATHIONE PEROXIDASE (E.C.1.11.1.9)	BOVINE (BOS $TAURUS) ERYTHROCYTE	2.0
330	1GPD	D-GYCERALDEHYDE-3-PHOSPHATE DEHYDROGENASE (E.C.1.2.1.12)D-GLYCERALDEHY	LOBSTER (HOMARUS AMERICANUS)	2.9
331	3GPD	D-GLYCERALDEHYDE-3-PHOSPHATE DEHYDROGENASE (E.C.1.2.1.12)	HUMAN (HOMO SAPIENS) MUSCLE	3.5
332	4GPD	APO-D-GYCERALDEHYDE-3-PHOSPHATE DEHYDROGENASE (E.C.1.2.1.12)	LOBSTER (HOMARUS $AMERICANUS)	2.8
333	3GRS	GLUTATHIONE REDUCTASE (E.C.1.6.4.2), OXIDIZED FORM (E)	HUMAN (HOMO $SAPIENS) ERYTHROCYTE	1.54
334	1HBS	HEMOGLOBIN S (DEOXY)	HUMAN (HOMO SAPIENS)	3.0
335	1HCO	HEMOGLOBIN (CARBONMONOXY)	HUMAN (HOMO SAPIENS)	2.7
336	2HCO	HEMOGLOBIN (CARBONMONOXY)	HUMAN (HOMO SAPIENS)	2.7
337	1HDS	HEMOGLOBIN (SICKLE CELL)	VIRGINIA WHITE-TAILED DEER (ODOCOILEUS VIRGINIANUS)	1.98
338	1HFL	IG*G1 FAB FRAGMENT (HY/HEL$-5) AND LYSOZYME (E.C.3.2.1.17) COMPLEX	/BALB(SLASH)C$ MOUSE (MUS $MUSCULUS) AND CHICKEN (GALLU	2.54
339	1HFM	IG*G1 FV FRAGMENT (HY/HEL$-10) (MODEL)	MOUSE (MUS $MUSCULUS)	N.A.
340	2HFM	IG*G1 FV FRAGMENT (HY/HEL$-10) AND LYSOZYME (E.C.3.2.1.17) COMPLEX (MO	MOUSE (MUS $MUSCULUS) AND HEN (GALLUS $GALLUS) EGG WHIT	N.A.
341	3HFM	IG*G1 FAB FRAGMENT (HY/HEL$-10) AND LYSOZYME (E.C.3.2.1.17) COMPLEX	MOUSE (MUS $MUSCULUS) AND CHICKEN (GALLUS $GALLUS)	3.0
342	2HHB	HEMOGLOBIN (DEOXY)	HUMAN (HOMO SAPIENS)	1.74
343	3HHB	HEMOGLOBIN (DEOXY)	HUMAN (HOMO SAPIENS)	1.74
344	4HHB	HEMOGLOBIN (DEOXY)	HUMAN (HOMO SAPIENS)	1.74
345	1HHO	HEMOGLOBIN A (OXY)	HUMAN (HOMO SAPIENS)	2.1
346	1HIP	OXIDIZED HIGH POTENTIAL IRON PROTEIN (HIPIP).	(CHROMATIUM VINOSUM), STRAIN D	2.0
347	2HIR	HIRUDIN (WILD-TYPE) (/NMR$,32 SIMULATED ANNEALING STRUCTURES)	LEECH (HIRUDO $MEDICINALIS) RECOMBINANT IN (ESCHERICHIA	N.A.
348	4HIR	HIRUDIN (MUTANT WITH LYS 47 REPLACED BY GLU) (/K47E$) (/NMR$,32 SIMULA	LEECH (HIRUDO $MEDICINALIS) RECOMBINANT IN (ESCHERICHIA	N.A.
349	5HIR	HIRUDIN (WILD-TYPE)	LEECH (HIRUDO $MEDICINALIS) RECOMBINANT IN (ESCHERICHIA	N.A.
350	6HIR	HIRUDIN (MUTANT WITH LYS 47 REPLACED BY GLU) (/K47E$) (/NMR$, MINIMIZE	LEECH (HIRUDO $MEDICINALIS) RECOMBINANT IN (ESCHERICHIA	N.A.

PDB	CODE	COMPOUND	SOURCE	RESOLUTION
351	1HKG	HEXOKINASE A AND GLUCOSE COMPLEX (E.C.2.7.1.1)	YEAST (SACCHAROMYCES CEREVISAE)	3.5
352	1HLA	HUMAN CLASS I HISTOCOMPATIBILITY ANTIGEN A2 (/HLA-A2$, HUMAN LEUCOCYTE	HUMAN (HOMO $SAPIENS) LYMPHOBLASTOID CELL LINE /JY$	3.5
353	2HLA	HUMAN CLASS I HISTOCOMPATIBILITY ANTIGEN AW 68.1 (/HLA-AW 68.1$, HUMA	HUMAN (HOMO $SAPIENS) LYMPHOBLASTOID CELL LINE /LB$	2.6
354	3HLA	HUMAN CLASS I HISTOCOMPATIBILITY ANTIGEN A2.1 (/HLA-A2.1$ HUMAN LEUCOC	HUMAN (HOMO $SAPIENS) LYMPHOBLASTOID CELL LINE /JY$	2.6
355	1HMG	HAEMAGGLUTININ (BROMELAIN DIGESTED)	INFLUENZA VIRUS (1968 X.31 STRAIN, V3 MUTANT)	3.0
356	1HMQ	HEMERYTHRIN (MET)	SIPUNCULID WORM (THEMISTE $DYSCRITUM)	2.0
357	1HMZ	HEMERYTHRIN (AZIDO, MET)	SIPUNCULID WORM (THEMISTE $DYSCRITUM)	2.0
358	1HNE	HUMAN NEUTROPHIL ELASTASE (/HNE$) (E.C.3.4.21.37) (ALSO REF KETONE (/M	HUMAN (HOMO $SAPIENS) NEUTROPHILS ISOLATED FROM PURULEN	1.84
359	1HOE	ALPHA-*AMYLASE INHIBITOR HOE-467*A	(STREPTOMYCES $TENDAE 4158)	5.5
360	1HR3	HEMERYTHRIN (AZIDO,MET)	(SIPHONOSOMA SPECIES NEAR SIPHONOSOMA $FUNAFATI)	5.5
361	1HRB	HEMERYTHRIN B	MARINE WORM (PHASCOLOPSIS GOULDII, SYNONYM GOLFINGIA GO	N.A.
362	1HVP	/HIV$-1 PROTEASE COMPLEX WITH SUBSTRATE (MODEL)	/HIV$-1 RETROVIRUS (STRAIN BH10)	N.A.
363	2HVP	/HIV$-1 PROTEASE	ESCHERICHIA $COLI (IN WHICH THE PROTEASE ENCODING GENE	3.0
364	3HVP	/(ABA$=6,7,95==)-/HIV$-1 PROTEASE (/SF2$ ISOLATE)	SYNTHETIC ENZYME CORRESPONDING TO THE PROTEASE FROM THE	2.8
365	4HVP	/HIV$-1S /PR$) COMPLEX WITH INHIBITOR N-AC $AMIDE (/M	SYNTHETIC ENZYME CORRESPONDING TO THE PROTEASE FROM THE	2.3
366	1HYA	HYALURONIC ACID (POLY D-GLUCURONIC ACID-N-ACETYL-D-GLUCOSAM SODIUM SAL	SYNTHESIZED BY EXTRACT OF RAT FIBROSARCOMA	3.0
367	2HYA	HYALURONIC ACID (POLY D-GLUCURONIC ACID-N-ACETYL-D-GLUCOSAM SODIUM SAL	HUMAN (HOMO SAPIENS) UMBILICAL CORD	3.0
368	3HYA	HYALURONIC ACID (POLY D-GLUCURONIC ACID-N-ACETYL-D-GLUCOSAM SODIUM SAL	HUMAN (HOMO SAPIENS) UMBILICAL CORD	3.0
369	4HYA	HYALURONIC ACID (POLY D-GLUCURONIC ACID-N-ACETYL-D-GLUCOSAM HIGH HUMID	HUMAN (HOMO SAPIENS) UMBILICAL CORD	3.0
370	1I1B	INTERLEUKIN-1*BETA (/IL$-1*BETA)	HUMAN (HOMO $SAPIENS) $SK-HEP-1 HEMATOMA CELLS, RECOMBI	2.0
371	2I1B	INTERLEUKIN-1*BETA (/IL$-1*BETA)	HUMAN (HOMO $SAPIENS) RECOMBINANT FORM EXPRESSED IN (ES	2.0
372	4I1B	INTERLEUKIN-1*BETA (/IL$-1*BETA)	HUMAN (HOMO $SAPIENS) RECOMBINANT FORM EXPRESSED IN (ES	2.0
373	3ICB	CALCIUM-BINDING PROTEIN (VITAMIN D-DEPENDENT, MINOR A FORM) (/ICABP$)	BOVINE (BOS $TAURUS) INTESTINE	2.3
374	2IG2	IMMUNOGLOBULIN G1	HUMAN (HOMO $SAPIENS) MYELOMA PATIENT KOL SERUM	N.A.
375	1IGE	FC FRAGMENT (IG*E(PRIME)/CL$) (MODEL)		N.A.
376	1INS	INSULIN	PIG (SUS SCROFA)	1.5
377	2INS	DES-*PHE B1 INSULIN	BOVINE (BOS $TAURUS)	2.5
378	3INS	2*ZN-INSULIN (JOINT X-RAY AND NEUTRON REFINEMENT)	PIG (SUS $SCROFA)	1.5
379	4INS	INSULIN	PIG (SUS $SCROFA)	1.5
380	2KAI	KALLIKREIN A (E.C.3.4.21.8)/BOVINE PANCREATIC TRYPSIN INHIBITOR COMPLE	PORCINE (SUS $SCROFA) PANCREAS AND BOVINE (BOS $TAURUS)	2.5
381	1KES	KERATAN SULFATE (SULFATED POLY(GALACTOSYL-N-ACETYL GLUCOSAMINE))	BOVINE (BOS TAURUS) CORNEA	3.0
382	1KGA	2-KETO-3-DEOXY-6-PHOSPHOGLUCONATE (/KDPG$) ALDOLASE (E.C.4.1.2.14)	(PSEUDOMONAS PUTIDA)	3.5
383	1L01	LYSOZYME (E.C.3.2.1.17) (DOUBLE MUTANT WITH THR 155 REPLACED BY ALA AN	BACTERIOPHAGE T4 (MUTANT GENE IS DERIVED FROM THE M13 P	1.7
384	1L02	LYSOZYME (E.C.3.2.1.17) (MUTANT WITH THR 157 REPLACED BY ALA) (T157A)	BACTERIOPHAGE T4 (MUTANT GENE IS DERIVED FROM THE M13 P	1.7
385	1L03	S==GAMMA157==-BETA-MERCAPTOETHANOL-LYSOZYME (E.C.3.2.1.17) (MUTANT WIT	BACTERIOPHAGE T4 (MUTANT GENE IS DERIVED FROM THE M13 P	1.7
386	1L04	LYSOZYME (E.C.3.2.1.17) (MUTANT WITH THR 157 REPLACED BY ASP) (T157D)	BACTERIOPHAGE T4 (MUTANT GENE IS DERIVED FROM THE M13 P	1.7
387	1L05	LYSOZYME (E.C.3.2.1.17) (MUTANT WITH THR 157 REPLACED BY GLU) (T157E)	BACTERIOPHAGE T4 (MUTANT GENE IS DERIVED FROM THE M13 P	1.7
388	1L06	LYSOZYME (E.C.3.2.1.17) (MUTANT WITH THR 157 REPLACED BY PHE) (T157F)	BACTERIOPHAGE T4 (MUTANT GENE IS DERIVED FROM THE M13 P	1.7
389	1L07	LYSOZYME (E.C.3.2.1.17) (MUTANT WITH THR 157 REPLACED BY GLY) (T157G)	BACTERIOPHAGE T4 (MUTANT GENE IS DERIVED FROM THE M13 P	1.7
390	1L08	LYSOZYME (E.C.3.2.1.17) (MUTANT WITH THR 157 REPLACED BY HIS) (T157H)	BACTERIOPHAGE T4 (MUTANT GENE IS DERIVED FROM THE M13 P	1.7
391	1L09	LYSOZYME (E.C.3.2.1.17) (MUTANT WITH THR 157 REPLACED BY ILE) (T157I)	BACTERIOPHAGE T4 (MUTANT GENE IS DERIVED FROM THE M13 P	1.7
392	1L10	S==GAMMA97==-BETA-MERCAPTOETHANOL-LYSOZYME (E.C.3.2.1.17) (MUTANT WITH	BACTERIOPHAGE T4 (MUTANT GENE IS DERIVED FROM THE M13 P	1.7
393	1L11	LYSOZYME (E.C.3.2.1.17) (MUTANT WITH THR 157 REPLACED BY ASN) (T157N)	BACTERIOPHAGE T4 (MUTANT GENE IS DERIVED FROM THE M13 P	1.7
394	1L12	LYSOZYME (E.C.3.2.1.17) (MUTANT WITH THR 157 REPLACED BY ARG) (T157R)	BACTERIOPHAGE T4 (MUTANT GENE IS DERIVED FROM THE M13 P	1.7
395	1L13	LYSOZYME (E.C.3.2.1.17) (MUTANT WITH THR 157 REPLACED BY SER) (T157S)	BACTERIOPHAGE T4 (MUTANT GENE IS DERIVED FROM THE M13 P	1.7
396	1L14	LYSOZYME ,(E.C.3.2.1.17) (MUTANT WITH THR 157 REPLACED BY SER) (T157S)	BACTERIOPHAGE T4 (MUTANT GENE IS DERIVED FROM THE M13 P	1.7
397	1L15	LYSOZYME (E.C.3.2.1.17) (MUTANT WITH THR 157 REPLACED BY VAL) (T157V)	BACTERIOPHAGE T4 (MUTANT GENE IS DERIVED FROM THE M13 P	1.7
398	1L16	LYSOZYME (E.C.3.2.1.17) (MUTANT WITH GLY 156 REPLACED BY ASP) (G156D)	BACTERIOPHAGE T4 (MUTANT GENE IS DERIVED FROM THE M13 P	1.7
399	1L17	LYSOZYME (E.C.3.2.1.17) (MUTANT WITH ILE 3 REPLACED BY VAL) (/I3V$)	BACTERIOPHAGE T4 (MUTANT GENE IS DERIVED FROM THE M13 P	1.7
400	1L18	LYSOZYME (E.C.3.2.1.17) (MUTANT WITH ILE 3 REPLACED BY TYR) (/I3Y$)	BACTERIOPHAGE T4 (MUTANT GENE IS DERIVED FROM THE M13 P	1.7

PDB CODE		COMPOUND	SOURCE	RESOLUTION
401	1L19	LYSOZYME (E.C.3.2.1.17) (MUTANT WITH SER 38 REPLACED BY ASP) (/S38D$)	BACTERIOPHAGE T4 (MUTANT GENE IS DERIVED FROM THE M13 P	1.7
402	1L20	LYSOZYME (E.C.3.2.1.17) (MUTANT WITH ASN 144 REPLACED BY ASP) (/N144D$)	BACTERIOPHAGE T4 (MUTANT GENE IS DERIVED FROM THE M13 P	1.85
403	1L21	LYSOZYME (E.C.3.2.1.17) (MUTANT WITH ASN 55 REPLACED BY GLY) (/N55G$)	BACTERIOPHAGE T4 (MUTANT GENE IS DERIVED FROM THE M13 P	1.85
404	1L22	LYSOZYME (E.C.3.2.1.17) (MUTANT WITH LYS 124 REPLACED BY GLY) (/K124G$)	BACTERIOPHAGE T4 (MUTANT GENE IS DERIVED FROM THE M13 P	1.7
405	1L23	LYSOZYME (E.C.3.2.1.17) (MUTANT WITH GLY 77 REPLACED BY ALA) (/G77A$)	BACTERIOPHAGE T4 (MUTANT GENE IS DERIVED FROM THE M13 P	1.7
406	1L24	LYSOZYME (E.C.3.2.1.17) (MUTANT WITH ALA 82 REPLACED BY PRO) (/A82P$)	BACTERIOPHAGE T4 (MUTANT GENE IS DERIVED FROM THE M13 P	1.7
407	1L25	LYSOZYME (E.C.3.2.1.17) (MUTANT WITH PRO 86 REPLACED BY ALA) (/P86A$)	BACTERIOPHAGE T4 (MUTANT GENE IS DERIVED FROM THE M13 P	1.7
408	1L26	S==GAMMA86==-BETA-MERCAPTOETHANOL-LYSOZYME (E.C.3.2.1.17) (MUTANT WITH	BACTERIOPHAGE T4 (MUTANT GENE IS DERIVED FROM THE M13 P	1.7
409	1L27	LYSOZYME (E.C.3.2.1.17) (MUTANT WITH PRO 86 REPLACED BY ASP) (/P86D$)	BACTERIOPHAGE T4 (MUTANT GENE IS DERIVED FROM THE M13 P	1.7
410	1L28	LYSOZYME (E.C.3.2.1.17) (MUTANT WITH PRO 86 REPLACED BY GLY) (/P86G$)	BACTERIOPHAGE T4 (MUTANT GENE IS DERIVED FROM THE M13 P	1.9
411	1L29	LYSOZYME (E.C.3.2.1.17) (MUTANT WITH PRO 86 REPLACED BY HIS) (/P86H$)	BACTERIOPHAGE T4 (MUTANT GENE IS DERIVED FROM THE M13 P	1.7
412	1L30	LYSOZYME (E.C.3.2.1.17) (MUTANT WITH PRO 86 REPLACED BY LEU) (/P86L$)	BACTERIOPHAGE T4 (MUTANT GENE IS DERIVED FROM THE M13 P	1.8
413	1L31	LYSOZYME (E.C.3.2.1.17) (MUTANT WITH PRO 86 REPLACED BY ARG) (/P86R$)	BACTERIOPHAGE T4 (MUTANT GENE IS DERIVED FROM THE M13 P	1.8
414	1L32	LYSOZYME (E.C.3.2.1.17) (MUTANT WITH PRO 86 REPLACED BY SER) (/P86S$)	BACTERIOPHAGE T4 (MUTANT GENE IS DERIVED FROM THE M13 P	1.7
415	1L33	LYSOZYME (E.C.3.2.1.17) (MUTANT WITH VAL 131 REPLACED BY ALA) (/V131A$)	BACTERIOPHAGE T4 (MUTANT GENE IS DERIVED FROM THE M13 P	1.7
416	1L34	LYSOZYME (E.C.3.2.1.17) (MUTANT WITH ARG 96 REPLACED BY HIS) (/R96H$)	BACTERIOPHAGE T4 (MUTANT GENE IS DERIVED FROM THE M13 P	1.7
417	1L35	LYSOZYME (E.C.3.2.1.17) (MUTANT WITH ILE 3 REPLACED BY TYR, (/C54T$, C	BACTERIOPHAGE T4 (MUTANT GENE IS DERIVED FROM THE M13 P	1.8
418	2LBP	LEUCINE-BINDING PROTEIN (/LBP$)	(ESCHERICHIA $COLI) STRAIN K12	2.4
419	1LDB	APO-*L-*LACTATE DEHYDROGENASE (E.C.1.1.1.27)	(BACILLUS $STEAROTHERMOPHILUS)	2.8
420	2LDB	L-*LACTATE DEHYDROGENASE (E.C.1.1.1.27) COMPLEX WITH /NAD$ AND FRUCTOS	(BACILLUS $STEAROTHERMOPHILUS)	3.0
421	3LDH	LACTATE DEHYDROGENASE (E.C.1.1.1.27) M4 COMPLEX, TERNARY COMPLEX WITH /	DOGFISH (SQUALUS ACANTHIUS) MUSCLE	3.0
422	5LDH	LACTATE DEHYDROGENASE H4= AND S-SLAC-/NAD$===+== COMPLEX (E.C.1.1.1.27	PIG (SUS SCROFA) HEART	2.7
423	6LDH	M=4= APO-*LACTATE DEHYDROGENASE (E.C.1.1.1.27)	DOGFISH (SQUALUS $ACANTHIAS) MUSCLE	2.8
424	8LDH	M=4= APO-*LACTATE DEHYDROGENASE (E.C.1.1.1.27) COMPLEX WITH CITRATE	DOGFISH (SQUALUS $ACANTHIAS) MUSCLE	2.8
425	1LDM	M=4= LACTATE DEHYDROGENASE (E.C.1.1.1.27) TERNARY COMPLEX WITH /NAD$ A	DOGFISH (SQUALUS $ACANTHIAS) MUSCLE	2.1
426	2LDX	APO-LACTATE DEHYDROGENASE (E.C.1.1.1.27), ISOENZYME C=4=	MOUSE (MUS $MUSCULUS) TESTICLES, SWISS-*WEBSTER STRAIN	2.96
427	1LH1	LEGHEMOGLOBIN (ACETATE,MET)	YELLOW LUPIN (LUPINUS LUTEUS L) ROOT NODULES	2.0
428	2LH1	LEGHEMOGLOBIN (ACETATE,MET)	YELLOW LUPIN (LUPINUS LUTEUS L) ROOT NODULES	2.0
429	1LH2	LEGHEMOGLOBIN (AQUO,MET)	YELLOW LUPIN (LUPINUS LUTEUS L) ROOT NODULES	2.0
430	2LH2	LEGHEMOGLOBIN (AQUO,MET)	YELLOW LUPIN (LUPINUS LUTEUS L) ROOT NODULES	2.0
431	1LH3	LEGHEMOGLOBIN (CYANO,MET)	YELLOW LUPIN (LUPINUS LUTEUS L) ROOT NODULES	2.0
432	2LH3	LEGHEMOGLOBIN (CYANO,MET)	YELLOW LUPIN (LUPINUS LUTEUS L) ROOT NODULES	2.0
433	1LH4	LEGHEMOGLOBIN (DEOXY)	YELLOW LUPIN (LUPINUS LUTEUS L) ROOT NODULES	2.0
434	2LH4	LEGHEMOGLOBIN (DEOXY)	YELLOW LUPIN (LUPINUS LUTEUS L) ROOT NODULES	2.0
435	1LH5	LEGHEMOGLOBIN (FLUORO,MET)	YELLOW LUPIN (LUPINUS LUTEUS L) ROOT NODULES	2.0
436	2LH5	LEGHEMOGLOBIN (FLUORO,MET)	YELLOW LUPIN (LUPINUS LUTEUS L) ROOT NODULES	2.0
437	1LH6	LEGHEMOGLOBIN (NICOTINATE,MET)	YELLOW LUPIN (LUPINUS LUTEUS L) ROOT NODULES	2.0
438	2LH6	LEGHEMOGLOBIN (NICOTINATE,MET)	YELLOW LUPIN (LUPINUS LUTEUS L) ROOT NODULES	2.0
439	1LH7	LEGHEMOGLOBIN (NITROSOBENZENE)	YELLOW LUPIN (LUPINUS LUTEUS L) ROOT NODULES	2.0
440	2LH7	LEGHEMOGLOBIN (NITROSOBENZENE)	YELLOW LUPIN (LUPINUS LUTEUS L) ROOT NODULES	2.0
441	2LHB	HEMOGLOBIN V (CYANO,MET)	SEA LAMPREY (PETROMYZON $MARINUS)	2.4
442	2LIV	LEUCINE(SLASH)*ISOLEUCINE(SLASH)*VALINE-BINDING PROTEIN (/LIVBP$)	(ESCHERICHIA $COLI)	3.0
443	1LLC	L-*LACTATE DEHYDROGENASE (E.C.1.1.1.27) COMPLEX WITH FRUCTOSE-1,6-BISP	(LACTOBACILLUS $CASEI)	2.5
444	1LRD	LAMBDA REPRESSOR-OPERATOR COMPLEX	BACTERIOPHAGE (LAMBDA)	3.20
445	1LRP	LAMBDA REPRESSOR (N-TERMINAL DOMAIN)	(LAMBDA)	
446	1LYD	T4*LYSOZYME	SYNTHETIC CODING /DNA$ EXPRESSED IN (ESCHERICHIA $COLI)	
447	1LYM	LYSOZYME (E.C.3.2.1.17)	HEN (GALLUS GALLUS) EGG WHITE	2.5
448	2LYM	LYSOZYME (E.C.3.2.1.17) (1 ATMOSPHERE, 1.4 M NA*CL)	HEN (GALLUS $GALLUS) EGG WHITE	2.0
449	3LYM	LYSOZYME (E.C.3.2.1.17) (1000 ATMOSPHERES, 1.4 M NA*CL)	HEN (GALLUS $GALLUS) EGG WHITE	2.0
450	1LYZ	LYSOZYME (E.C.3.2.1.17)	HEN (GALLUS GALLUS) EGG WHITE	2.0

219

PDB CODE	COMPOUND	SOURCE	RESOLUTION
451 2LYZ	LYSOZYME (E.C.3.2.1.17)	HEN (GALLUS GALLUS) EGG WHITE	2.0
452 3LYZ	LYSOZYME (E.C.3.2.1.17)	HEN (GALLUS GALLUS) EGG WHITE	2.0
453 4LYZ	LYSOZYME (E.C.3.2.1.17)	HEN (GALLUS GALLUS) EGG WHITE	2.0
454 5LYZ	LYSOZYME (E.C.3.2.1.17)	HEN (GALLUS GALLUS) EGG WHITE	2.0
455 6LYZ	LYSOZYME (E.C.3.2.1.17)	HEN (GALLUS GALLUS) EGG WHITE	2.0
456 7LYZ	LYSOZYME (E.C.3.2.1.17) TRICLINIC CRYSTAL FORM	HEN (GALLUS GALLUS) EGG WHITE	2.5
457 8LYZ	LYSOZYME (E.C.3.2.1.17) IODINE-INACTIVATED	HEN (GALLUS GALLUS) EGG WHITE	2.5
458 9LYZ	LYSOZYME (NAM-NAG-NAM SUBSTRATE ONLY) (E.C.3.2.1.17)	HEN (GALLUS GALLUS) EGG-WHITE	1.5
459 1LZ1	LYSOZYME (E.C.3.2.1.17)	HUMAN (HOMO $SAPIENS)	2.8
460 1LZ2	LYSOZYME (E.C.3.2.1.17)	TURKEY (MELEAGRIS GALLOPAVO) EGG WHITE	2.2
461 2LZ2	LYSOZYME (E.C.3.2.1.17)	TURKEY (MELEAGRIS $GALLOPAVO) EGG WHITE	2.2
462 1LZH	LYSOZYME (MONOCLINIC) (E.C.3.2.1.17)	HEN (GALLUS GALLUS) EGG WHITE	6.0
463 2LZH	LYSOZYME (ORTHORHOMBIC) (E.C.3.2.1.17)	HEN (GALLUS GALLUS) EGG WHITE	6.0
464 1LZM	LYSOZYME (E.C.3.2.1.17)	(ESCHERICHIA $COLI INFECTED WITH BACTERIOPHAGE T4)	1.7
465 3LZM	LYSOZYME (E.C.3.2.1.17)	(ESCHERICHIA $COLI INFECTED WITH BACTERIOPHAGE T4)	1.7
466 1LZT	LYSOZYME (E.C.3.2.1.17), TRICLINIC CRYSTAL FORM	HEN (GALLUS $GALLUS) EGG WHITE	1.97
467 2LZT	LYSOZYME (E.C.3.2.1.17), TRICLINIC CRYSTAL FORM	HEN (GALLUS $GALLUS) EGG WHITE	1.97
468 1MB5	MYOGLOBIN (CARBONMONOXYMYOGLOBIN) (NEUTRON STUDY)	SPERM WHALE (PHYSETER $CATODON)	1.8
469 1MBA	MYOGLOBIN (MET) ($P*H 7.0)	SEA HARE (APLYSIA $LIMACINA)	1.6
470 2MBA	MYOGLOBIN (AZIDE COMPLEX) ($P*H 7.0)	SEA HARE (APLYSIA $LIMACINA)	1.6
471 3MBA	MYOGLOBIN (FLUORIDE COMPLEX) ($P*H 7.0)	SEA HARE (APLYSIA $LIMACINA)	2.0
472 4MBA	MYOGLOBIN (IMIDAZOLE COMPLEX) ($P*H 7.0)	SEA HARE (APLYSIA $LIMACINA)	2.0
473 1MBC	MYOGLOBIN (FE /IIS, CARBONMONOXY, 260 DEGREES K)	SPERM WHALE (PHYSETER $CATODON)	1.5
474 1MBD	MYOGLOBIN (DEOXY, $P*H 8.4)	SPERM WHALE (PHYSETER CATODON)	1.4
475 1MBN	MYOGLOBIN (FERRIC IRON - METMYOGLOBIN)	SPERM WHALE (PHYSETER CATODON)	2.0
476 4MBN	MYOGLOBIN (MET)	SPERM WHALE (PHYSETER $CATODON)	2.0
477 5MBN	MYOGLOBIN (DEOXY)	SPERM WHALE (PHYSETER CATODON)	2.0
478 1MBO	MYOGLOBIN (OXY, $P*H 8.4)	SPERM WHALE (PHYSETER CATODON)	1.6
479 1MBS	MYOGLOBIN (MET)	COMMON SEAL (PHOCA VITULINA)	2.5
480 1MCG	IMMUNOGLOBULIN, LAMBDA-*TYPE BENCE-*JONES DIMER MCG	HUMAN (HOMO SAPIENS)	2.3
481 1MCP	IMMUNOGLOBULIN FAB FRAGMENT (MC/PC5603)	MOUSE (MUS $MUSCULUS)	2.7
482 2MCP	IMMUNOGLOBULIN MC/PC603$ FAB-PHOSPHOCHOLINE COMPLEX	MOUSE (MUS $MUSCULUS)	3.1
483 4MDH	CYTOPLASMIC MALATE DEHYDROGENASE (E.C.1.1.1.37)	PORCINE (SUS $SCROFA) HEART	2.5
484 2MEV	MENGO ENCEPHALOMYOCARDITIS VIRUS COAT PROTEIN	MONKEY BRAIN,MIDDLE SIZE PLAQUE VARIANT,MENGO VIRUS	3.0
485 2MHB	HEMOGLOBIN (HORSE, AQUO MET)	HORSE (EQUUS CABALLUS)	2.0
486 2MHR	MYOHEMERYTHRIN	SIPUNCULAN WORM (THEMISTE $ZOSTERICOLA) RETRACTOR MUSCL	1.7+
487 1MLP	MUREIN LIPOPROTEIN	(ESCHERICHIA COLI)	N.A.
488 1MLT	MELITTIN	HONEY BEE (APIS MELLIFERA) VENOM	2.0
489 1MON	MONELLIN	SERENDIPITY (DIOSCOREOPHYLLUM $CUMMINSII DIELS) BERRY	2.75
490 1NTP	MODIFIED BETA TRYPSIN (MONOISOPROPYLPHOSPHORYL INHIBITED) (E.C.3.4.21.	BOVINE (BOS $TAURUS) PANCREAS	1.8
491 1NXB	NEUROTOXIN $B (PROBABLY IDENTICAL TO ERABUTOXIN $B)	SEA SNAKE (LATICAUDA SEMIFASCIATA) FROM PHILIPPINES SEA	1.38
492 1OVO	OVOMUCOID THIRD DOMAIN	JAPANESE QUAIL (COTURNIX COTURNIX JAPONICA)	1.9
493 2OVO	OVOMUCOID THIRD DOMAIN	SILVER PHEASANT (LOPHURA $NYCTHEMERA)	1.5
494 1P01	ALPHA-LYTIC PROTEASE (E.C.3.4.21.12) COMPLEX WITH BOC-*ALA-*PRO-*VALIN	(LYSOBACTER $ENZYMOGENES 495)	2.0
495 1P02	ALPHA-LYTIC PROTEASE (E.C.3.4.21.12) COMPLEX WITH METHOXYSUCCINYL-*ALA	(LYSOBACTER $ENZYMOGENES 495)	2.0
496 1P03	ALPHA-LYTIC PROTEASE (E.C.3.4.21.12) COMPLEX WITH METHOXYSUCCINYL-*ALA	(LYSOBACTER $ENZYMOGENES 495)	2.15
497 1P04	ALPHA-LYTIC PROTEASE (E.C.3.4.21.12) COMPLEX WITH METHOXYSUCCINYL-*ALA	(LYSOBACTER $ENZYMOGENES 495)	2.55
498 1P05	ALPHA-LYTIC PROTEASE (E.C.3.4.21.12) COMPLEX WITH METHOXYSUCCINYL-*ALA	(LYSOBACTER $ENZYMOGENES 495)	2.10
499 1P06	ALPHA-LYTIC PROTEASE (E.C.3.4.21.12) COMPLEX WITH METHOXYSUCCINYL-*ALA	(LYSOBACTER $ENZYMOGENES 495)	2.34
500 1P07	ALPHA-LYTIC PROTEASE (E.C.3.4.21.12) (MUTANT WITH MET 192 REPLACED BY	(LYSOBACTER $ENZYMOGENES 495)	2.25

	PDB CODE	COMPOUND	SOURCE	RESOLUTION
501	1P08	ALPHA-LYTIC PROTEASE (E.C.3.4.21.12) (MUTANT WITH MET 192 R *ALA-*ALA-	(LYSOBACTER $ENZYMOGENES 495)	2.25
502	1P09	ALPHA-LYTIC PROTEASE (E.C.3.4.21.12) (MUTANT WITH MET 213 REPLACED BY	(LYSOBACTER $ENZYMOGENES 495)	2.20
503	1P10	ALPHA-LYTIC PROTEASE (E.C.3.4.21.12) (MUTANT WITH MET 213 *ALA-*ALA-*P	(LYSOBACTER $ENZYMOGENES 495)	2.25
504	2P21	$C-*H-RAS $P21 PROTEIN CATALYTIC DOMAIN	TRANSFORMED (ESCHERICHIA $COLI) HARBORING A PLASMID THA	2.2
505	3P21	$C-*H-RAS $P21 PROTEIN CATALYTIC DOMAIN (MUTANT WITH GLY 12 REPLACED B	TRANSFORMED (ESCHERICHIA $COLI) HARBORING A PLASMID THA	2.2
506	1P2P	PHOSPHOLIPASE A=2= (E.C.3.1.1.4) (PHOSPHATIDE ACYL-HYDROLASE)	PORCINE (SUS $CROFA) PANCREAS	2.6
507	3P2P	PHOSPHOLIPASE A=2= (PHOSPHATIDE-2-ACYL-HYDROLASE) MUTANT WI (/D59S$, /	PORCINE (SUS $SCROFA) PANCREAS	2.1
508	2PAB	PREALBUMIN (HUMAN PLASMA)	HUMAN (HOMO SAPIENS)	2.1
509	1PAD	PAPAIN (E.C.3.4.22.2) -ACETYL-ALANYL-ALANYL- PHENYLALANYL-M CYSTEINE 2	PAPAYA (CARICA PAPAYA) FRUIT LATEX	2.8
510	2PAD	PAPAIN (E.C.3.4.22.2) -CYSTEINYL DERIVATIVE OF CYSTEINE-25 (/PAPSSCYS)	PAPAYA (CARICA PAPAYA) FRUIT LATEX	2.8
511	4PAD	PAPAIN (E.C.3.4.22.2) -TOSYL-METHYLENYLLYSYL DERIVATIVE OF CYSTEINE-25	PAPAYA (CARICA PAPAYA) FRUIT LATEX	2.8
512	5PAD	PAPAIN (E.C.3.4.22.2) -BENZYLOXYCARBONYL-GLYCYL- PHENYLALAN (/ZGPGCK)	PAPAYA (CARICA PAPAYA) FRUIT LATEX	2.8
513	6PAD	PAPAIN (E.C.3.4.22.2) -BENZYLOXYCARBONYL- PHENYLALANYL-METHYLENYLALANY	PAPAYA (CARICA PAPAYA) FRUIT LATEX	1.65
514	9PAP	PAPAIN (E.C.3.4.22.2) CYS-25 OXIDIZED	PAPAYA (CARICA $PAPAYA) FRUIT LATEX	1.55
515	1PAZ	PSEUDOAZURIN (OXIDIZED CU ++ AT $P*H 6.8)	(ALCALIGENES $FAECALIS, STRAIN S-6)	1.65
516	2PAZ	PSEUDOAZURIN (CUPREDOXIN)	(ALCALIGENES $FAECALIS, STRAIN S-6)	1.55
517	1PCY	PLASTOCYANIN (CU ++, $P*H 6.0)	POPLAR (POPULUS NIGRA VARIANT ITALICA) LEAVES	1.6
518	2PCY	APO-PLASTOCYANIN ($P*H 6.0)	POPLAR (POPULUS $NIGRA VARIANT ITALICA) LEAVES	1.8
519	3PCY	PLASTOCYANIN (HG==2=== SUBSTITUTED)	POPLAR (POPULUS $NIGRA VARIANT ITALICA) LEAVES	1.9
520	4PCY	PLASTOCYANIN (CROSS-LINKED WITH GLUTERALDEHYDE, CU+1, $P*H 7.8)	POPLAR (POPULUS $NIGRA VARIANT ITALICA) LEAVES	2.15
521	5PCY	PLASTOCYANIN (CU+1,$P*H 7.0)	POPLAR (POPULUS $NIGRA VARIANT ITALICA) LEAVES	1.80
522	6PCY	PLASTOCYANIN (CU+1,$P*H 3.8)	POPLAR (POPULUS $NIGRA VARIANT ITALICA) LEAVES	1.90
523	1PEP	PEPSIN (E.C.3.4.23.1)	PIG (SUS $CROFA)	3.0
524	2PEP	PEPSIN (E.C.3.4.23.1)	PORCINE (SUS $SCROFA)	2.3
525	3PEP	PEPSIN (E.C.3.4.23.1)	PIG (SUS $SCROFA)	2.3
526	4PEP	PEPSIN (E.C.3.4.23.1)	PIG (SUS $SCROFA)	1.8
527	1PFC	P/FC(PRIME) FRAGMENT OF AN IG*G1	GUINEA PIG (CAVIA $PORCELLUS) SERUM	3.12
528	1PFK	PHOSPHOFRUCTOKINASE (E.C.2.7.1.11) (R-STATE) COMPLEX WITH FRUCTOSE-1, 6	(ESCHERICHIA $COLI)	2.4
529	2PFK	PHOSPHOFRUCTOKINASE (E.C.2.7.1.11)	(ESCHERICHIA $COLI)	2.4
530	3PFK	PHOSPHOFRUCTOKINASE (E.C.2.7.1.11)	(BACILLUS $STEAROTHERMOPHILUS)	2.4
531	4PFK	PHOSPHOFRUCTOKINASE (E.C.2.7.1.11) COMPLEX WITH FRUCTOSE-6-PHOSPHATE A	(BACILLUS $STEAROTHERMOPHILUS)	2.4
532	5PFK	PHOSPHOFRUCTOKINASE (E.C.2.7.1.11) (INHIBITED T-STATE CRYSTALLIZED WIT	(BACILLUS $STEAROTHERMOPHILUS)	7.0
533	1PGI	D-GLUCOSE 6-PHOSPHATE ISOMERASE (E.C.5.3.1.9)	PORCINE (SUS $CROFA) MUSCLE	3.5
534	2PGK	PHOSPHOGLYCERATE KINASE (HORSE, MUSCLE) (E.C.2.7.2.3)	HORSE (EQUUS CABALLUS) MUSCLE	3.0
535	3PGK	PHOSPHOGLYCERATE KINASE (E.C.2.7.2.3) COMPLEX WITH ATP, MAGNESIUM OR M	BAKERS YEAST (SACCHAROMYCES CEREVISIAE)	2.5
536	3PGM	PHOSPHOGLYCERATE MUTASE (E.C.2.7.5.3) DE-PHOSPHO ENZYMEPHOSPHOGLYCEROM	DRIED BAKER,S YEAST (SACCHAROMYCES CEREVISIAE)	2.8
537	1PHH	$P-*HYDROXYBENZOATE HYDROXYLASE (/PHBH$) (E.C.1.14.13.2) - (/PHBH$./FA	PSEUDOMONAS $FLUORESCENS)	2.3
538	1PHY	PHOTOACTIVE YELLOW PROTEIN	(ECTOTHIORHODOSPIRA $HALOPHILA)	2.4
539	2PKA	KALLIKREIN A (E.C.3.4.21.8)	PORCINE (SUS $SCROFA) PANCREAS	2.05
540	2PLV	POLIOVIRUS (TYPE 1, MAHONEY STRAIN)	HUMAN (HOMO $SAPIENS) POLIOVIRUS (TYPE 1, MAHONEY STRAI	2.88
541	1PMB	MYOGLOBIN (AQUOMET, $P*H 7.1)	PORCINE (SUS $SCROFA) RECOMBINANT FROM (ESCHERICHIA $CO	2.5
542	1PP2	CALCIUM-FREE PHOSPHOLIPASE A=2= (E.C.3.1.1.4)	WESTERN DIAMONDBACK RATTLESNAKE (CROTALUS $ATROX)	2.5
543	1PPD	2-HYDROXYETHYLTHIOPAPAIN (E.C.3.4.22.22)- CRYSTAL FORM D	PAPAYA (CARICA $PAPAYA) FRUIT LATEX	2.0
544	1PPT	AVIAN PANCREATIC POLYPEPTIDE	TURKEY (MELEAGRIS GALLOPAVO) PANCREAS	1.37
545	1PRC	PHOTOSYNTHETIC REACTION CENTER	(RHODOPSEUDOMONAS $VIRIDIS)	2.3
546	2PRK	PROTEINASE K (E.C.3.4.21.14)	FUNGUS (TRITIRACHIUM $ALBUM LIMBER)	1.5
547	1PSG	PEPSINOGEN	PORCINE (SUS $SCROFA)	1.65
548	3PTB	BETA-TRYPSIN (BENZAMIDINE INHIBITED) AT $P*H7 (E.C.3.4.21.4	BOVINE (BOS TAURUS) PANCREAS	1.7
549	2PTC	BETA-TRYPSIN (E.C.3.4.21.4) COMPLEX WITH PANCREATIC TRYPSIN INHIBITOR	BOVINE (BOS TAURUS) PANCREAS	1.9
550	1PTE	D-ALANYL-*D-ALANINE CARBOXYPEPTIDASE(SLASH)TRANSPEPTIDASE	(STREPTOMYCES R61)	2.8

PDB	CODE	COMPOUND	SOURCE	RESOLUTION
551	4PTI	TRYPSIN INHIBITOR (CRYSTAL FORM /II$)	BOVINE (BOS $TAURUS) PANCREAS	1.5
552	5PTI	TRYPSIN INHIBITOR (CRYSTAL FORM /II$)	BOVINE (BOS $TAURUS) PANCREAS	1.0
553	6PTI	BOVINE PANCREATIC TRYPSIN INHIBITOR (/BPTI$, CRYSTAL FORM /III$)	BOVINE (BOS $TAURUS) PANCREAS	1.7
554	2PTN	TRYPSIN (ORTHORHOMBIC, 2.4 M AMMONIUM SULFATE) (E.C.3.4.21.4)	BOVINE (BOS $TAURUS) PANCREAS	1.55
555	3PTN	TRYPSIN (TRIGONAL, 2.4 M AMMONIUM SULFATE) (E.C.3.4.21.4)	BOVINE (BOS $TAURUS) PANCREAS	1.7
556	4PTP	BETA TRYPSIN, DIISOPROPYLPHOSPHORYL INHIBITED (E.C.3.4.21.4)	BOVINE (BOS $TAURUS) PANCREAS	1.34
557	1PYK	PYRUVATE KINASE (E.C.2.7.1.40)	CAT MUSCLE (FELIS DOMESTICA)	2.6
558	1PYP	INORGANIC PYROPHOSPHATASE (E.C.3.6.1.1)	BAKER'S YEAST (SACCHAROMYCES CEREVISIAE)	3.0
559	2R04	RHINOVIRUS 14 (/HRV$14) COMPLEX WITH ANTIVIRAL AGENT /WIN IV$	HUMAN (HOMO $SAPIENS) VIRUS GROWN IN HE*LA CELLS	3.0
560	2R06	RHINOVIRUS 14 (/HRV$14) COMPLEX WITH ANTIVIRAL AGENT /WIN VI$	HUMAN (HOMO $SAPIENS) VIRUS GROWN IN HE*LA CELLS	3.0
561	2R07	RHINOVIRUS 14 (/HRV$14) COMPLEX WITH ANTIVIRAL AGENT /WIN VII$	HUMAN (HOMO $SAPIENS) VIRUS GROWN IN HE*LA CELLS	3.0
562	1R08	RHINOVIRUS 14 (/HRV$14) COMPLEX WITH ANTIVIRAL AGENT /WIN VIII$	HUMAN (HOMO $SAPIENS) VIRUS GROWN IN HE*LA CELLS	3.0
563	1R69	434 REPRESSOR (AMINO-TERMINAL DOMAIN) (R1-69)	PHAGE 434	2.0
564	1RBB	RIBONUCLEASE B(E.C.3.1.4.22)	BOVINE (BOS $TAURUS) PANCREAS	2.5
565	1RDG	RUBREDOXIN	(DESULFOVIBRIO $GIGAS)	1.4
566	1REI	BENCE-*JONES IMMUNOGLOBULIN /REI$ VARIABLE PORTION	HUMAN (HOMO SAPIENS)	2.0
567	1RHD	RHODANESE (E.C.2.8.1.1)	BOVINE (BOS $TAURUS) LIVER	2.5
568	2RHE	BENCE-*JONES PROTEIN (LAMBDA, VARIABLE DOMAIN)	HUMAN (HOMO SAPIENS) MYELOMA PATIENT RHE URINE	1.6
569	4RHV	RHINOVIRUS 14 (/HRV$14)	HUMAN (HOMO $SAPIENS) VIRUS GROWN IN HE*LA CELLS	3.0
570	1RLX	RELAXIN	PIG (SUS SCROFA) OVARY	N.A.
571	2RLX	RELAXIN	PIG (SUS SCROFA) OVARY	N.A.
572	3RLX	RELAXIN	PIG (SUS SCROFA) OVARY	N.A.
573	4RLX	RELAXIN	PIG (SUS SCROFA) OVARY	N.A.
574	2RM2	RHINOVIRUS 14 (/HRV$14) COMPLEX WITH ANTIVIRAL AGENT /WIN II(S/R)$	HUMAN (HOMO $SAPIENS) VIRUS GROWN IN HE*LA CELLS	3.0
575	1RMU	RHINOVIRUS 14 (/HRV$14) (MUTANT WITH CYS 1 199 REPLACED BY TYR) /C199	HUMAN (HOMO $SAPIENS) VIRUS GROWN IN HE*LA CELLS, MUTAN	3.0
576	2RMU	RHINOVIRUS 14 (/HRV$14) (MUTANT WITH VAL 1 188 REPLACED BY LEU) /V188	HUMAN (HOMO $SAPIENS) VIRUS GROWN IN HE*LA CELLS, MUTAN	3.0
577	1RN3	RIBONUCLEASE A (E.C.3.1.27.5)	BOVINE (BOS $TAURUS) PANCREAS	1.45
578	1RNS	RIBONUCLEASE-S (E.C.3.1.4.22)RIBONUCLEASE (EC 3.1.27.5), PANCREATIC	BOVINE (BOS $TAURUS) PANCREAS	2.0
579	1RNT	RIBONUCLEASE T=1=(E.C.3.1.27.3) ISOZYME-2 (PRIME)-GUANYLIC ACID COMPLEX	(ASPERGILLUS $ORYZAE)	1.9
580	2RNT	LYS 25-RIBONUCLEASE T=1= (LYS 25-/RN$ASE T=1=) (E.C.3.1.27. COMPLEX WI	(ASPERGILLUS $ORYZAE)	1.8
581	3RNT	LYS 25-RIBONUCLEASE T=1= (LYS 25-/RN$ASE T=1=) (E.C.3.1.27. COMPLEX WI	(ASPERGILLUS $ORYZAE)	1.8
582	3RP2	RAT MAST CELL PROTEASE /II$ (/RMCPII$)	RAT (RATTUS) ATYPICAL MAST CELLS FROM THE SMALL	1.9
583	1RS1	RHINOVIRUS 14 (/HRV$14) COMPLEX WITH ANTIVIRAL AGENT /WIN I(R)$	HUMAN (HOMO $SAPIENS) VIRUS GROWN IN HE*LA CELLS	1.9
584	2RS1	RHINOVIRUS 14 (/HRV$14) COMPLEX WITH ANTIVIRAL AGENT /WIN I(S)$	HUMAN (HOMO $SAPIENS) VIRUS GROWN IN HE*LA CELLS	3.0
585	2RS3	RHINOVIRUS 14 (/HRV$14) COMPLEX WITH ANTIVIRAL AGENT /WIN III(S)$	HUMAN (HOMO $SAPIENS) VIRUS GROWN IN HE*LA CELLS	3.0
586	2RS5	RHINOVIRUS 14 (/HRV$14) COMPLEX WITH ANTIVIRAL AGENT /WIN V(S)$	HUMAN (HOMO $SAPIENS) VIRUS GROWN IN HE*LA CELLS	3.0
587	5RSA	RIBONUCLEASE A (E.C.3.1.4.22) (JOINT NEUTRON AND X-RAY)	BOVINE (BOS $TAURUS) PANCREAS	2.0
588	6RSA	RIBONUCLEASE A (E.C.3.1.27.5) COMPLEX WITH URIDINE VANADATE (JOINT NEU	BOVINE (BOS $TAURUS) PANCREAS	2.0
589	7RSA	RIBONUCLEASE A (PHOSPHATE-FREE) (E.C.3.1.27.5)	BOVINE (BOS $TAURUS) PANCREAS	1.26
590	1RSM	LYS-7-(DINITROPHENYLENE)-LYS-41 CROSS-LINKED RIBONUCLEASE *A (E.C.3.1	BOVINE (BOS $TAURUS) PANCREAS	2.0
591	2RSP	ROUS SARCOMA VIRUS PROTEASE (/RSV PR$)	ROUS SARCOMA VIRUS (STRAIN PR-*C)	2.0
592	2RUB	RU*BIS*C*O (RIBULOSE-1,5-*BISPHOSPHATE CARBOXYLASE (SLASH) OXYGENASE) (E	(RHODOSPIRILLUM $RUBRUM)	1.7
593	3RXN	RUBREDOXIN	(DESULFOVIBRIO VULGARIS)	1.5
594	4RXN	RUBREDOXIN (OXIDIZED, FE(/III$)) (UNCONSTRAINED MODEL)	(CLOSTRIDIUM $PASTEURIANUM)	1.20
595	5RXN	RUBREDOXIN (OXIDIZED, FE(/III$)) (CONSTRAINED MODEL)	(CLOSTRIDIUM $PASTEURIANUM)	1.20
596	1SBC	SUBTILISIN CARLSBERG (SUBTILOPEPTIDASE *A) (E.C.3.4.21.14)	(BACILLUS $SUBTILIS)	2.5
597	1SBT	SUBTILISIN /BPN$* (E.C.3.4.21.14)	PROBABLY BACILLUS AMYLOLIQUEFACIENS	2.5
598	2SBT	SUBTILISIN NOVO (E.C.3.4.21.14)	PROBABLY BACILLUS AMYLOLIQUEFACIENS	2.8
599	4SBV	SOUTHERN BEAN MOSAIC VIRUS COAT PROTEIN	SOUTHERN BEAN MOSAIC VIRUS (COW PEA STRAIN)	2.8
600	2SEC	SUBTILISIN CARLSBERG (E.C.3.4.21.14) COMPLEX WITH GENETICALLY-ENGINEER	(BACILLUS $SUBTILIS) AND LEECH (HIRUDO $MEDICINALIS)	1.8

PDB CODE	COMPOUND	SOURCE	RESOLUTION
601 2SGA	PROTEINASE A (COMPONENT OF THE EXTRACELLULAR FILTRATE PRONASE) (/SGPA$)	(STREPTOMYCES $GRISEUS, STRAIN K1)	1.5
602 3SGB	PROTEINASE B FROM STREPTOMYCES GRISEUS (/SGPB$) (E.C. NUMBEPROTEASE B	(STREPTOMYCES $GRISEUS, STRAIN K1) AND TURKEY (MELEAGRI	1.8
603 1SGC	PROTEINASE *A COMPLEX WITH CHYMOSTATIN	(STREPTOMYCES $GRISEUS, STRAIN K1)	1.8
604 1SGT	TRYPSIN (/SGT$) (E.C.3.4.21.4)	(STREPTOMYCES $GRISEUS, STRAIN K1)	1.7
605 1SIC	SUBTILISIN /BPN$ /BPN$ (PRIME) (E.C.3.4.21.14) COMPLEX WITH STREPTOMYCES SUBT	PROBABLY (BACILLUS $AMYLOLIQUEFACIENS) AND (STREPTOMYCE	2.0
606 1SN3	SCORPION NEUROTOXIN (VARIANT 3)	SCORPION (CENTRUROIDES SCULPTURATUS EWING)	1.8
607 2SNI	SUBTILISIN NOVO (E.C.3.4.21.14) COMPLEX WITH CHYMOTRYPSIN INHIBITOR 2	(BACILLUS $AMYLOLIQUEFACIENS) AND BARLEY (HORDEUM $VULG	2.1
608 2SNS	STAPHYLOCOCCAL NUCLEASE (E.C.3.1.4.7) COMPLEX WITH 2(PRIME)STAPHYLOCOC	(STAPHYLOCOCCUS AUREUS), FOGGI STRAIN	1.5
609 2SOD	CU,ZN SUPEROXIDE DISMUTASE (E.C.1.6.4.5) (OXIDIZED FORM)	BOVINE (BOS TAURUS) ERYTHROCYTE	2.0
610 1SRX	THIOREDOXIN $COLI B)	(ESCHERICHIA $COLI B)	2.8
611 2SSI	STREPTOMYCES SUBTILISIN INHIBITOR	(STREPTOMYCES ALBOGRISEOLUS, S-3253)	2.6
612 2STV	SATELLITE TOBACCO NECROSIS VIRUS	COAT PROTEIN OF SATELLITE TOBACCO NECROSIS VIRUS	2.50
613 2TAA	TAKA-*AMYLASE A (E.C.3.2.1.1)	(ASPERGILLUS ORYZAE)	3.0
614 2BVV	TOMATO BUSHY STUNT VIRUS	TOMATO BUSHY STUNT VIRUS	2.90
615 1TEC	THERMITASE (E.C.3.4.21.14) COMPLEX WITH EGLIN-C	(THERMOACTINOMYCES $VULGARIS) AND LEECH (HIRUDO $MEDICI	2.2
616 2TGA	TRYPSINOGEN (2.4 M MAGNESIUM SULFATE)	BOVINE (BOS TAURUS) PANCREAS	1.8
617 1TGB	TRYPSINOGEN-CA FROM PEG	BOVINE (BOS TAURUS) PANCREAS	1.8
618 1TGC	TRYPSINOGEN (0.50 METHANOL, 0.50 WATER)	BOVINE (BOS TAURUS) PANCREAS	2.1
619 2TGD	TRYPSINOGEN, DIISOPROPYLPHOSPHORYL INHIBITED	BOVINE (BOS $TAURUS) PANCREAS	1.65
620 1TGN	TRYPSINOGEN	BOVINE (BOS TAURUS) PANCREAS	1.9
621 2TGP	TRYPSINOGEN COMPLEX WITH PANCREATIC TRYPSIN INHIBITOR	BOVINE (BOS TAURUS) PANCREAS	1.8
622 1TGS	TRYPSINOGEN COMPLEX WITH PORCINE PANCREATIC SECRETORY TRYPSIN INHIBITO	BOVINE (BOS TAURUS) PANCREAS AND PORCINE (SUS SCROFA) P	1.8
623 1TGT	TRYPSINOGEN (173 DEGREES K, 0.70 METHANOL, 0.30 WATER)	BOVINE (BOS TAURUS) PANCREAS	1.7
624 2TGT	TRYPSINOGEN (103 DEGREES K, 0.70 METHANOL, 0.30 WATER)	BOVINE (BOS TAURUS) PANCREAS	1.7
625 1THI	THAUMATIN I	KETEMFE (THAUMATOCOCCUS $DANIELLII BENTH) BERRY	3.2
626 1TIM	TRIOSE PHOSPHATE ISOMERASE (E.C.5.3.1.1)	CHICKEN (GALLUS GALLUS) BREAST MUSCLE	2.5
627 1TLD	BETA-TRYPSIN (ORTHORHOMBIC) AT $P*H 5.3 (E.C.3.4.21.4)	BOVINE (BOS $TAURUS) PANCREAS	1.5
628 3TLN	THERMOLYSIN (E.C.3.4.24.4)	(BACILLUS THERMOPROTEOLYTICUS)	1.6
629 4TLN	THERMOLYSIN (E.C.3.4.24.4) COMPLEX WITH L-LEUCYL-HYDROXYLAMINE	(BACILLUS THERMOPROTEOLYTICUS)	2.3
630 5TLN	THERMOLYSIN (E.C.3.4.24.4) COMPLEX WITH HONH-BENZYLMALONYL-L-ALANYLGLY	(BACILLUS THERMOPROTEOLYTICUS)	2.3
631 7TLN	THERMOLYSIN (E.C.3.4.24.4) COMPLEX WITH CH2CO(N-OH)LEU-OCH3	(BACILLUS THERMOPROTEOLYTICUS)	2.3
632 1TLP	THERMOLYSIN (E.C.3.4.24.4) COMPLEX WITH PHOSPHORAMIDON	(BACILLUS $THERMOPROTEOLYTICUS)	2.3
633 2TMA	TROPOMYOSIN	RABBIT (ORYCTOLAGUS $CUNICULUS) CARDIAC MUSCLE	15.0
634 1TMN	THERMOLYSIN (E.C.3.4.24.4) COMPLEX WITH N-(1-CARBOXY-3-PHENYLPROPYL)-*	(BACILLUS $THERMOPROTEOLYTICUS)	1.9
635 2TMN	THERMOLYSIN (E.C.3.4.24.4) COMPLEX WITH N-PHOSPHORYL-*L-LEUCINAMIDE (P	(BACILLUS $THERMOPROTEOLYTICUS)	1.6
636 3TMN	THERMOLYSIN (E.C.3.4.24.4) COMPLEX WITH VAL-*TRP (/VW$)	(BACILLUS $THERMOPROTEOLYTICUS)	1.7
637 4TMN	THERMOLYSIN (E.C.3.4.24.4) COMPLEX WITH CBZ-PHE==P==-LEU-AL (/ZFPLA$)	(BACILLUS $THERMOPROTEOLYTICUS)	1.7
638 5TMN	THERMOLYSIN (E.C.3.4.24.4) COMPLEX WITH CBZ-*GLY==P===*LEU-*LEU (Z*G==	(BACILLUS $THERMOPROTEOLYTICUS)	1.6
639 6TMN	THERMOLYSIN (E.C.3.4.24.4) COMPLEX WITH CBZ-GLY==P==-(O)-*LEU-*LEU (Z*	(BACILLUS $THERMOPROTEOLYTICUS)	1.6
640 7TMN	THERMOLYSIN (E.C.3.4.24.4) COMPLEX WITH GLY-*TPH-*LEU-*LEU (SUBSTRATE	(BACILLUS $THERMOPROTEOLYTICUS)	N.A.
641 2TMV	INTACT TOBACCO MOSAIC VIRUS (FIBER DIFFRACTION STUDY)	TOBACCO MOSAIC VIRUS ($VULGARE STRAIN)	2.9
642 1TN1	PB(//II$)-TRANSFER RIBO-NUCLEIC ACID (YEAST,PHE) T/RNA ($P*H 7.4)	YEAST (SACCHAROMYCES $CEREVISIAE)	3.0
643 1TN2	PB(//II$)-TRANSFER RIBO-NUCLEIC ACID (YEAST,PHE) T/RNA ($P*H 5.0)	YEAST (SACCHAROMYCES $CEREVISIAE)	3.0
644 4TNA	TRANSFER RIBO-NUCLEIC ACID (YEAST,PHE), $T/RNA	YEAST (SACCHAROMYCES CEREVISIAE)	2.5
645 6TNA	TRANSFER RIBO-NUCLEIC ACID (YEAST,PHE), $T/RNA	YEAST (SACCHAROMYCES CEREVISIAE)	2.7
646 1TNC	TROPONIN - CALCIUM-BINDING COMPONENT	RABBIT (ORYCTOLAGUS CUNICULUS)	N.A.
647 1TNC	TROPONIN-*C	CHICKEN (GALLUS $GALLUS) SKELETAL MUSCLE	2.0
648 5TNC	TROPONIN-*C	TURKEY (MELEAGRIS $GALLOPAVO) SKELETAL MUSCLE	2.0
649 1TNF	TUMOR NECROSIS FACTOR-ALPHA (CACHECTIN)	HUMAN (HOMO $SAPIENS) RECOMBINANT FORM EXPRESSED IN YEA	2.6
650 1TON	TONIN (E.C. NUMBER NOT ASSIGNED)	RAT (RATTUS $RATTUS) SUBMAXILLARY GLAND	1.8

PDB CODE	COMPOUND	SOURCE	RESOLUTION
651 1TPA	ANHYDRO-TRYPSIN (E.C.3.4.21.4) COMPLEX WITH PANCREATIC TRYPSIN INHIBIT	BOVINE (BOS TAURUS) PANCREAS	1.9
652 2TPI	TRYPSINOGEN - PANCREATIC TRYPSIN INHIBITOR - ILE-VAL COMPLE (2.4 M MAG	BOVINE (BOS TAURUS) PANCREAS	2.1
653 3TPI	TRYPSINOGEN COMPLEX WITH PANCREATIC TRYPSIN INHIBITOR AND ILE-VAL	BOVINE (BOS TAURUS) PANCREAS	1.9
654 4TPI	TRYPSINOGEN COMPLEX WITH THE ARG==15==ANALOGUE OF PANCREATIC TRYPSIN	BOVINE (BOS $TAURUS) PANCREAS	2.2
655 1TPO	BETA-TRYPSIN (ORTHORHOMBIC) AT SP*H5.0 (E.C.3.4.21.4)	BOVINE (BOS TAURUS) PANCREAS	1.7
656 1TPP	BETA-TRYPSIN (E.C.3.4.21.4) COMPLEX WITH P-AMIDINO-PHENYL-PYRUVATE (AP	BOVINE (BOS TAURUS) PANCREAS	1.4
657 1TRA	TRANSFER RIBO-NUCLEIC ACID (YEAST, PHE), TRNA	YEAST (SACCHAROMYCES $CEREVISIAE)	3.0
658 2TRA	TRANSFER RIBO-NUCLEIC ACID (YEAST, ASP) $T/RNA (*A FORM)	YEAST (SACCHAROMYCES $CEREVISIAE)	3.0
659 3TRA	TRANSFER RIBO-NUCLEIC ACID (YEAST, ASP) $T/RNA (B FORM)	YEAST (SACCHAROMYCES $CEREVISIAE)	3.0
660 4TRA	TRANSFER RIBO-NUCLEIC ACID (YEAST, PHE), $T/RNA	YEAST (SACCHAROMYCES $CEREVISIAE)	2.7
661 1TRM	ASN==102==*TRYPSIN (E.C.3.4.21.4) (MUTANT WITH ASP 102 REPL $P*H 6 (AN	RAT (RATTUS $RATTUS)	2.3
662 2TRM	ASN==102==*TRYPSIN (E.C.3.4.21.4) (MUTANT WITH ASP 102 REPL $P*H 8 (AN	RAT (RATTUS $RATTUS)	2.8
663 2TS1	TYROSYL-TRANSFER /RNA$ SYNTHETASE (E.C.6.1.1.1)	(BACILLUS $STEAROTHERMOPHILUS /NCA$ 1503)	2.3
664 3TS1	TYROSYL-TRANSFER /RNA$ SYNTHETASE (E.C.6.1.1.1) COMPLEXED WITH TYROSIN	(BACILLUS $STEAROTHERMOPHILUS /NCA$ 1503)	2.7
665 4TS1	DES-((ILE 318-ARG 417)-TYROSYL-TRANSFER /RNA$ SYNTHETASE (E. COMPLEXED	(BACILLUS $STEAROTHERMOPHILUS /NCA$ 1503) (MUTANT GENE	2.5
666 1UBQ	UBIQUITIN	HUMAN (HOMO $SAPIENS) ERYTHROCYTES	1.8
667 1UTG	UTEROGLOBIN (OXIDIZED)	RABBIT (ORYCTOLAGUS $CUNICULUS) FEMALE GENITAL TRACT	1.34
668 2UTG	UTEROGLOBIN	UTERINE SECRETIONS OF PREGNANT NEW ZEALAND RABBITS (ORY	1.64
669 3WGA	WHEAT GERM AGGLUTININ (ISOLECTIN 2)	WHEAT (TRITICUM $VULGARIS) GERM	1.8
670 1WRP	$TRP REPRESSOR (TRIGONAL FORM)	(ESCHERICHIA $COLI)	2.2
671 2WRP	$TRP REPRESSOR (ORTHORHOMBIC FORM)	(ESCHERICHIA $COLI)	1.65
672 3WRP	$TRP APOREPRESSOR	(ESCHERICHIA $COLI)	1.8
673 1WSY	TRYPTOPHAN SYNTHASE (E.C.4.2.1.20)	(SALMONELLA $TYPHIMURIUM, STRAIN /TB$2211(SLASH) $P/STH8	2.5
674 1XIA	D-*XYLOSE ISOMERASE (E.C.5.3.1.5)	(ARTHROBACTER, STRAIN R3728)	2.3
675 2XIA	D-*XYLOSE ISOMERASE (E.C.5.3.1.5)	(STREPTOMYCES $RUBIGINOSUS)	3.5
676 3XIA	D-*XYLOSE ISOMERASE (E.C.5.3.1.5)	(STREPTOMYCES $OLIVOCHROMOGENES)	3.0
677 4XIA	D-*XYLOSE ISOMERASE (E.C.5.3.1.5), D-*SORBITOL COMPLEX	(ARTHROBACTER, STRAIN B3728)	2.3
678 5XIA	D-*XYLOSE ISOMERASE (E.C.5.3.1.5), XYLITOL COMPLEX	(ARTHROBACTER, STRAIN B3728)	2.5
679 1XY1	1 BETA-MERCAPTOPROPIONATE-OXYTOCIN (WET FORM)	SYNTHETIC	2.3
680 1XY2	1 BETA-MERCAPTOPROPIONATE-OXYTOCIN (DRY FORM)	SYNTHETIC	1.04
681 2YHX	YEAST HEXOKINASE B (E.C.2.7.1.1) COMPLEX WITH2 ORTHO-TOLUOYLGLUCOSAMI	BAKERS YEAST (SACCHAROMYCES CEREVISIAE)	1.20
682 1ZNA	/DNA$ (Z*, 5*-$D(*CP*GP*CP*G)-3*, HIGH SALT)	SYNTHETIC DNA	1.6
683 2ZNA	/DNA$ (Z-I, 5*-$D(P*CP*GP*CP*GP*CP*G)-3*)	SYNTHETIC DNA	N.A.
684 3ZNA	/DNA$ (Z-II, 5*-$D(P*CP*GP*CP*GP*CP*GP*CP*G)-3*	SYNTHETIC DNA	N.A.

A 'ready reference' atlas of protein folding patterns

The reader of this book will realize by now that no single picture can express everything of significance about a protein. No single set of pictures can show everything about the known protein structures. The pictures here are limited to portraying the three-dimensional folding patterns of the backbones of proteins. In many cases only monomers are shown.

To facilitate the use of these pictures for reference, they appear in order of Data Bank code, ignoring the initial numeral. An extremely tempting alternative was to group them by structure type. But of course, readers may cut out and reassemble these pictures according to their own interests.

I myself have spent many enjoyable moments browsing through these pictures—tracing the chain here, picking out secondary and super-secondary structures there. I find that these structures, considered purely as abstract spatial patterns, have an attractiveness independent of the scientific importance of the corpus of data they represent and am continually reminded of Goethe's remark that 'Architecture is frozen music'. I hope that a substantial number of readers will share this feeling.

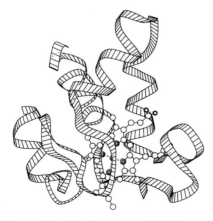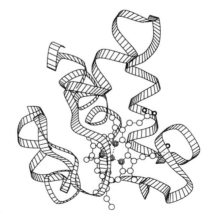

Figure A1. Cytochrome c_{551} (451C).

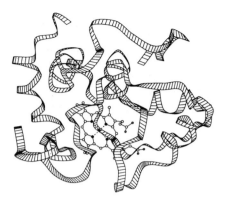
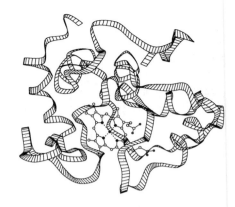

Figure A2. Cytochrome c_{550} (155C).

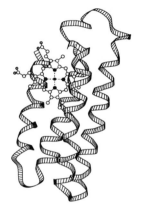
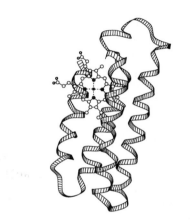

Figure A3. Cytochrome b_{562} (156B).

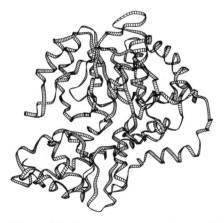
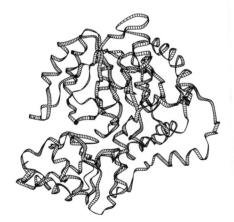

Figure A4. Aspartate aminotransferase (1AAT).

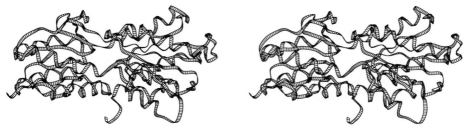

Figure A5. L-Arabinose binding protein (1ABP).

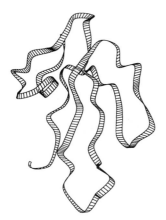
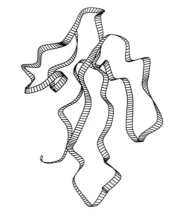

Figure A6. α-bungarotoxin (2ABX).

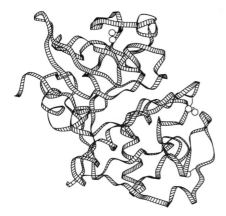
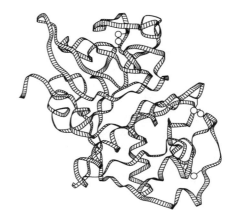

Figure A7. Actinidin (2ACT).

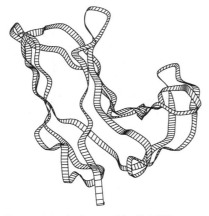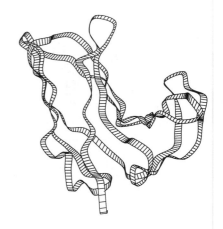

Figure A8. Actinoxanthin (1ACX).

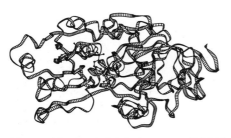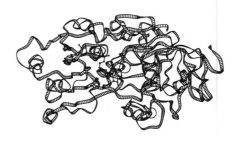

Figure A9. Alcohol dehydrogenase (6ADH).

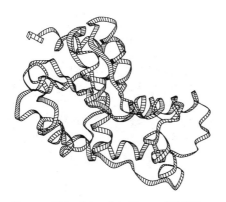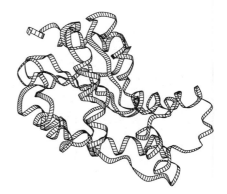

Figure A10. Adenylate kinase (3ADK).

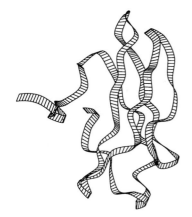 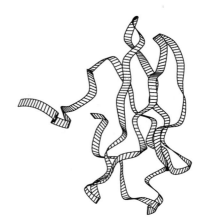

Figure A11. Tendimistat (2AIT).

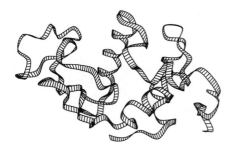 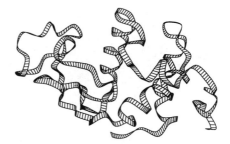

Figure A12. α-lactalbumin (1ALC).

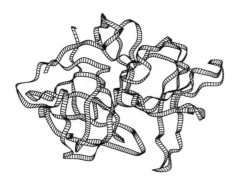 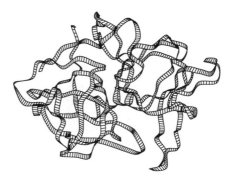

Figure A13. α-lytic protease (2ALP).

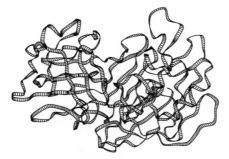
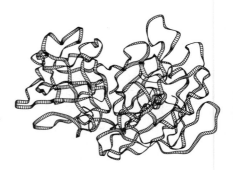

Figure A14. Endothiapepsin (4APE).

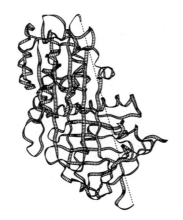
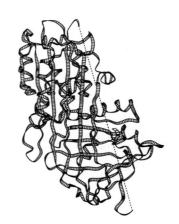

Figure A15. α_1-antitrypsin (5API).

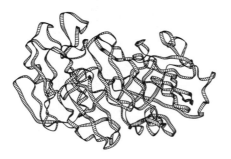
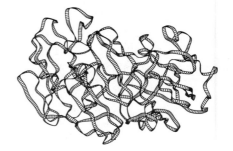

Figure A16. Penicillopepsin (2APP).

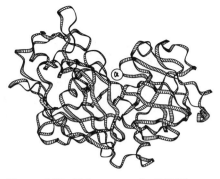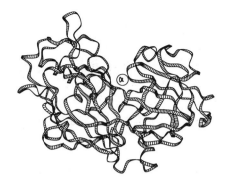

Figure A17. Rhizopuspepsin (2APR).

Figure A18. Aspartate transcarbamylase (4ATC).

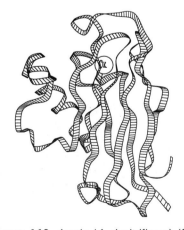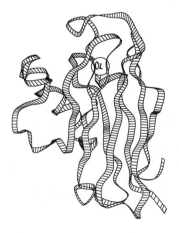

Figure A19. Azurin (*A. denitrificans*) (2AZA).

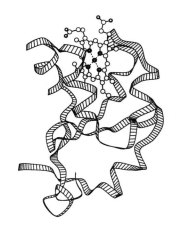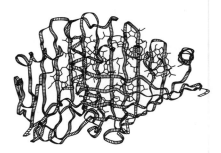

Figure A20. Cytochrome b_5 (2B5C).

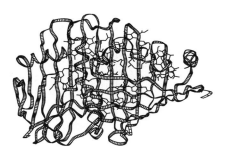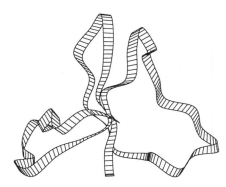

Figure A21. Bacteriochlorophyll (3BCL).

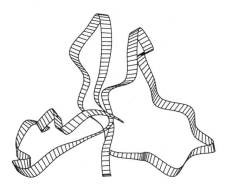

Figure A22. Sea anemone antiviral protein (1BDS).

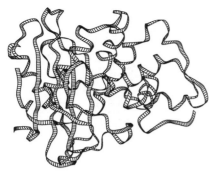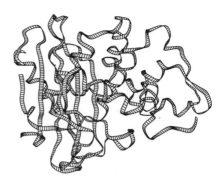

Figure A23. *β*-Lactamase (1BLM).

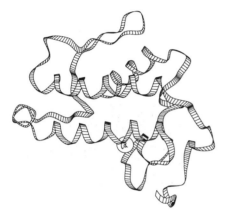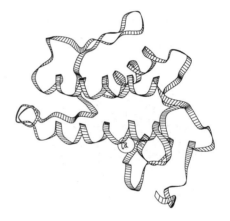

Figure A24. Phospholipase A$_2$ (1BP2).

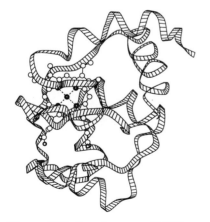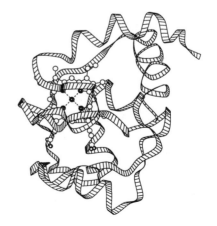

Figure A25. Cytochrome c_2 (3C2C).

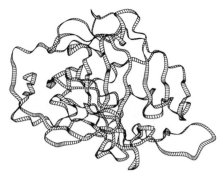
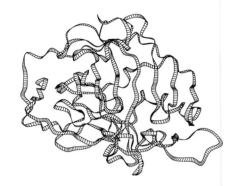

Figure A26. Carbonic anhydrase (2CA2).

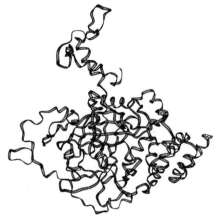
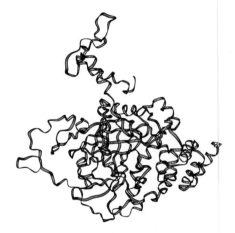

Figure A27. Beef liver catalase (8CAT).

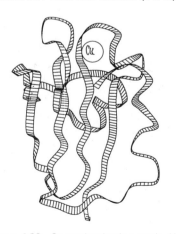
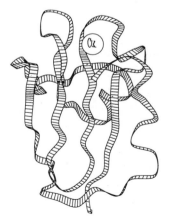

Figure A28. Cucumber basic protein (1CBP).

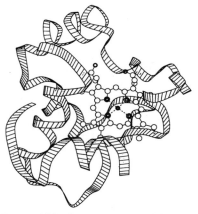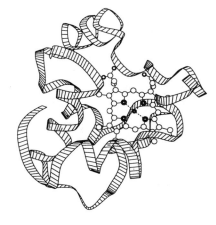

Figure A29. Cytochrome c_5 (1CC5).

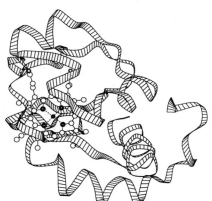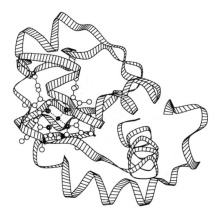

Figure A30. Rice cytochrome c (1CCR).

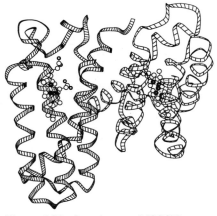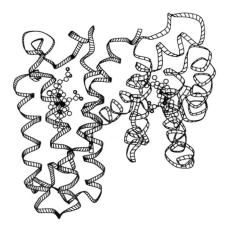

Figure A31. Cytochrome c' (2CCY).

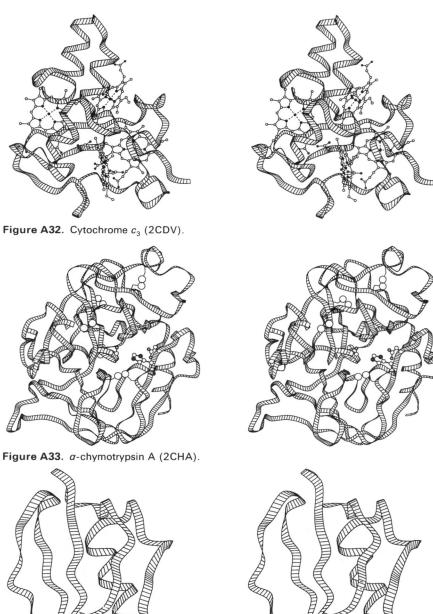

Figure A32. Cytochrome c_3 (2CDV).

Figure A33. a-chymotrypsin A (2CHA).

Figure A34. Barley chymotrypsin inhibitor (2CI2).

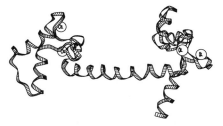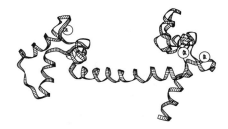

Figure A35. Calmodulin (3CLN).

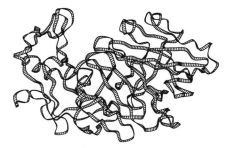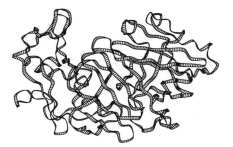

Figure A36. Chymosin B (1CMS).

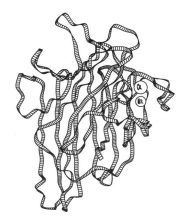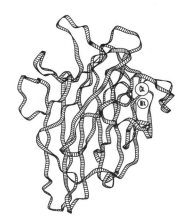

Figure A37. Concanavalin A (2CNA).

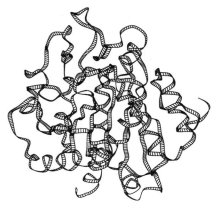
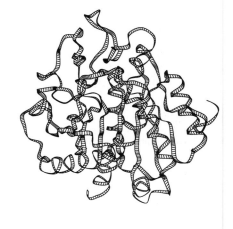

Figure A38. Carboxypeptidase A (5CPA).

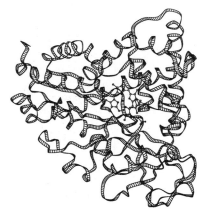
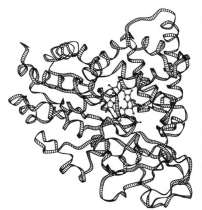

Figure A39. Cytochrome P450 CAM (2CPP).

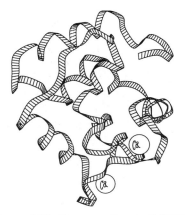
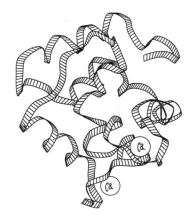

Figure A40. Carp parvalbumin (1CPV).

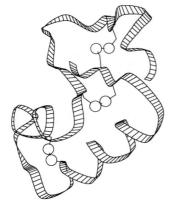
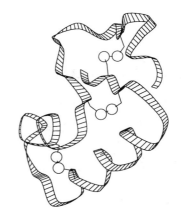

Figure A41. Crambin (1CRN).

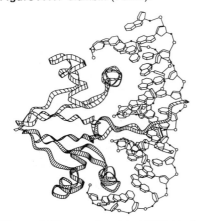
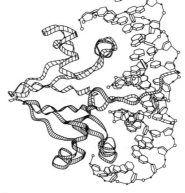

Figure A42. λ cro (model of DNA binding) (1CRO).

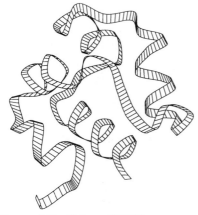
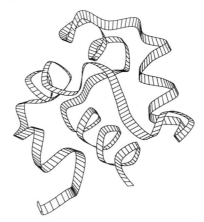

Figure A43. 434 cro (2CRO).

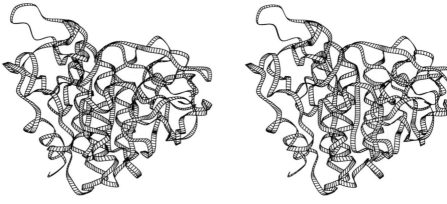

Figure A44. Subtilisin Carlsberg (1CSE).

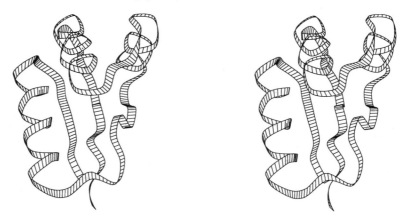

Figure A45. C-terminal domain of ribosomal protein L7L12 (1CTF).

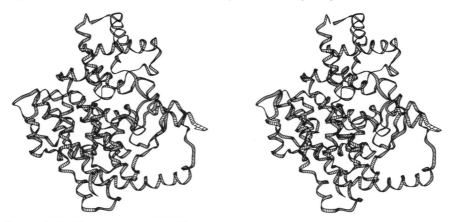

Figure A46. Citrate synthase (3CTS).

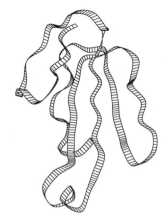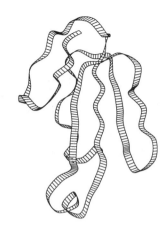

Figure A47. *α*-cobratoxin (1CTX).

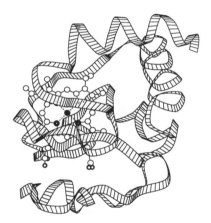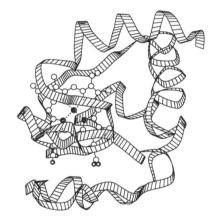

Figure A48. Tuna cytochrome *c* (5CYT).

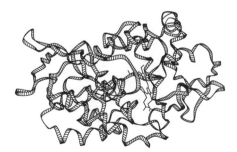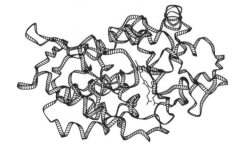

Figure A49. Cytochrome *c* peroxidase (2CYP).

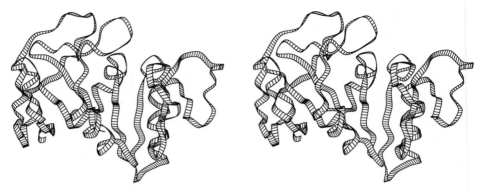

Figure A50. Dihydrofolate reductase (3DFR).

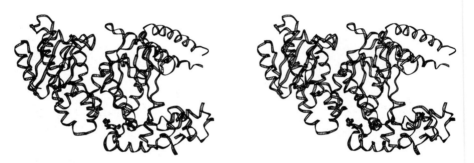

Figure A51. DNA polymerase I (Klenow fragment) (1DPI).

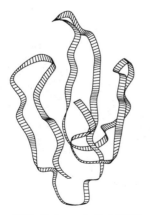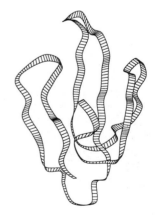

Figure A52. Erabutoxin B (5EBX).

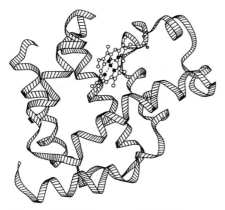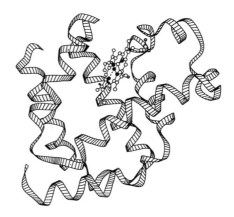

Figure A53. Erythrocruorin (1ECD).

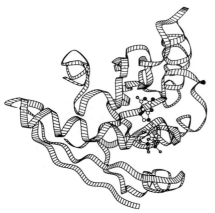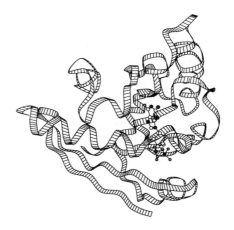

Figure A54. Elongation factor TU (1EFM).

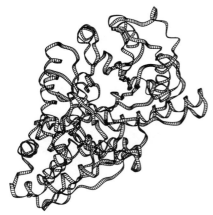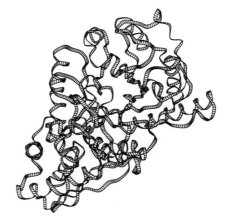

Figure A55. Enolase (2ENL).

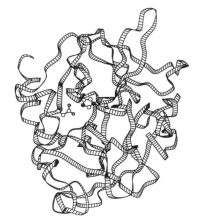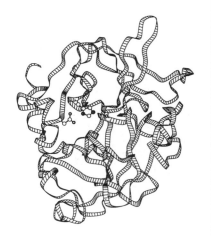

Figure A56. Porcine elastase (2EST).

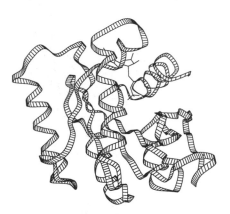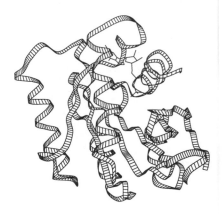

Figure A57. Elongation factor TU (1ETU).

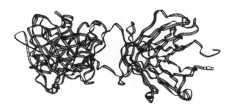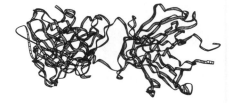

Figure A58. Fab KOL (2FB4).

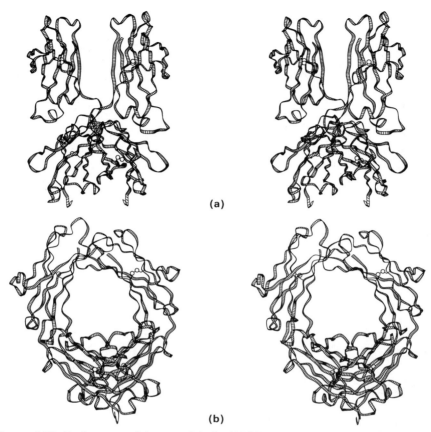

(a)

(b)

Figure A59. Fc fragment of immunoglobulin (1FC1). Parts (a) and (b) show different orientations.

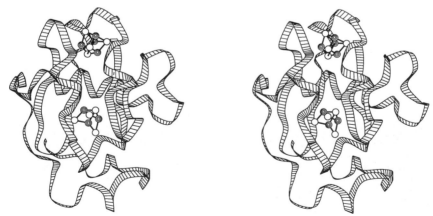

Figure A60. Ferredoxin (4FD1).

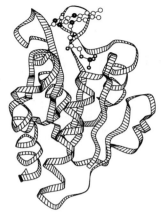 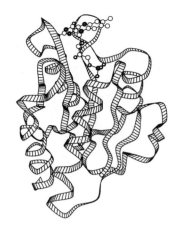

Figure A61. Flavodoxin (3FXN).

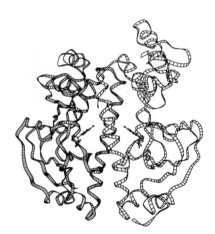 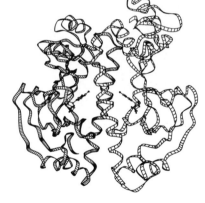

Figure A62. Catabolite gene activator protein (3GAP).

 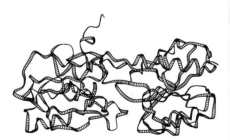

Figure A63. Galactose-binding protein (1GBP).

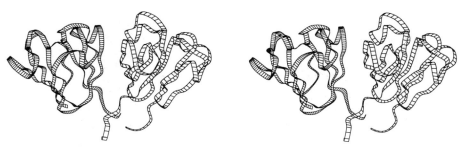

Figure A64. γ-crystallin (1GCR).

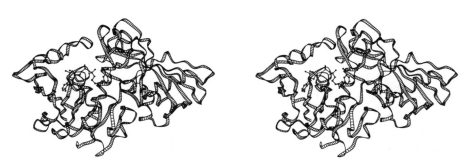

Figure A65. Glyceraldehyde-3-phosphate dehydrogenase (1GD1).

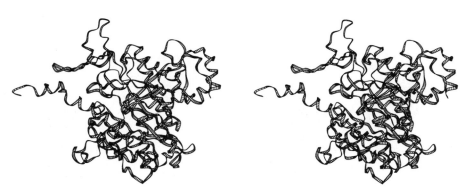

Figure A66. Glutamine synthetase (1GLS).

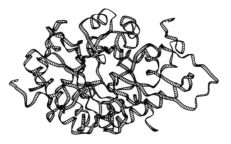
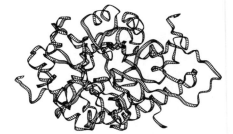

Figure A67. Spinach glycolate oxidase (1GOX).

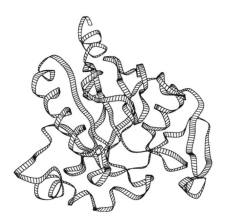

Figure A68. Glutathione peroxidase (1GP1).

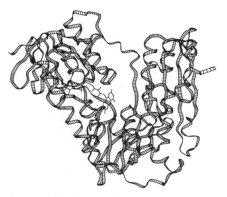
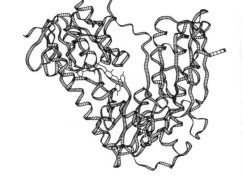

Figure A69. Glutathione reductase (3GRS).

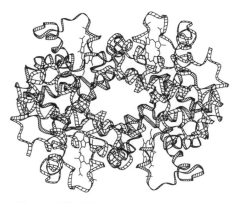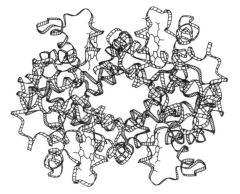

Figure A70. Human haemoglobin (1HHO).

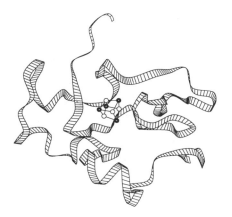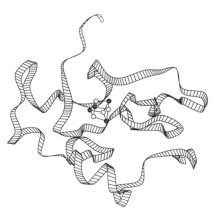

Figure A71. High potential iron protein (1HIP).

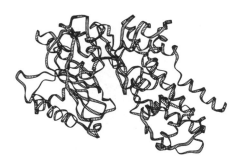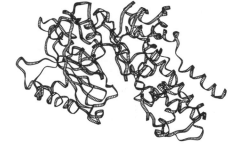

Figure A72. Hexokinase (1HKG).

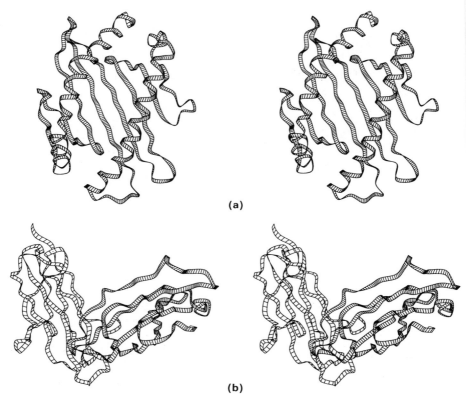

(a)

(b)

Figure A73. Human class I histocompatibility antigen (2HLA) (a) Domains α_1 and α_2; this unit contains the binding site for processed antigens: the groove between the two long helices. (b) α_3 and β_2 microglobulin.

Figure A74. Influenza haemagglutinin (1HMG).

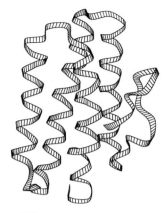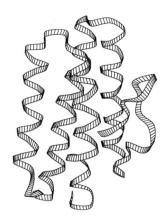

Figure A75. Haemerythrin (1HMQ).

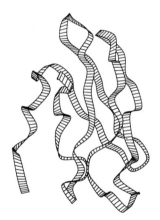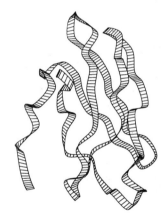

Figure A76. *a*-amylase inhibitor (1HOE).

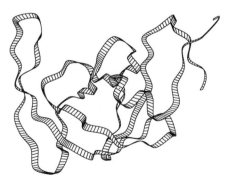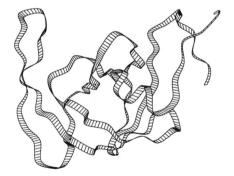

Figure A77. HIV Protease (3HVP).

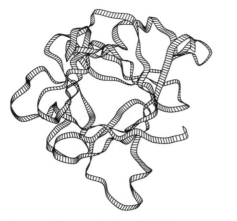

Figure A78. Interleukin-1β (2I1B).

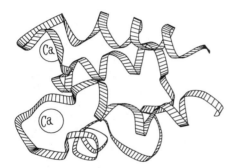
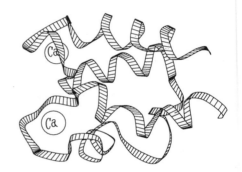

Figure A79. Intestinal calcium-binding protein (3ICB).

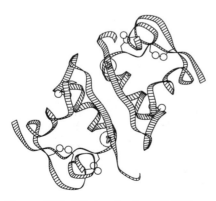
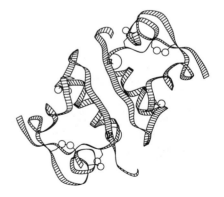

Figure A80. 2Zn Insulin dimer (1INS).

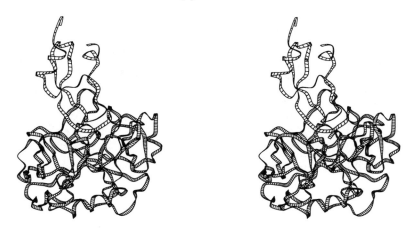

Figure A81. Kallekrein-Pancreatic Trypsin Inhibitor complex (2KAI).

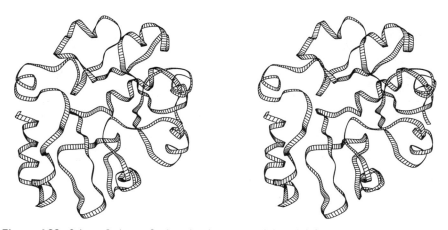

Figure A82. 2-keto-3-deoxy-6-phosphogluconate aldolase (1KGA).

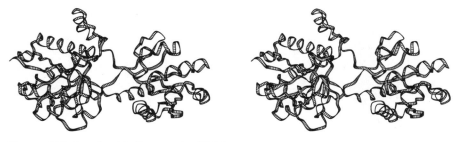

Figure A83. Leucine-binding protein (2LBP).

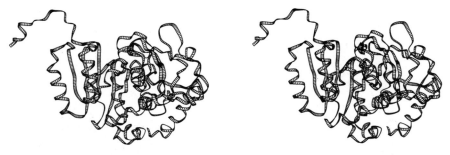

Figure A84. Dogfish lactate dehydrogenase (3LDH).

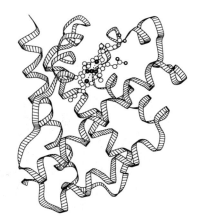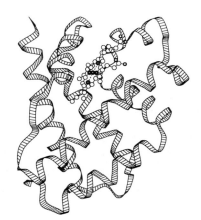

Figure A85. Lupin leghaemoglobin (2LH4).

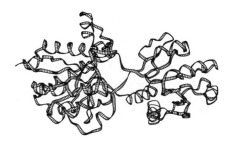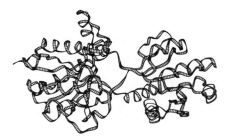

Figure A86. Leu/Ile/Val-binding protein (2LIV).

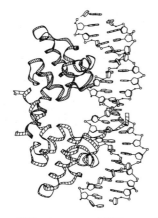
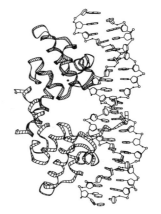

Figure A87. λ repressor/DNA complex (1LRD).

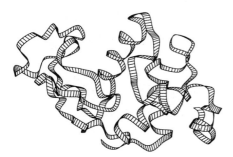
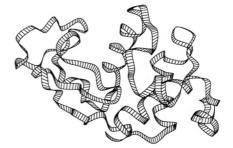

Figure A88. Hen egg white lysozyme (1LYZ).

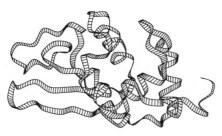
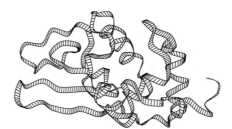

Figure A89. Human lysozyme (1LZ1).

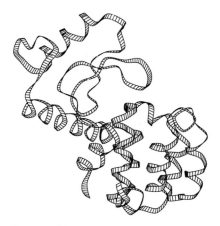

Figure A90. T4 lysozyme (3LZM).

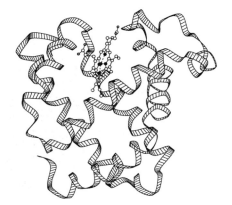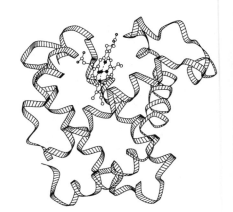

Figure A91. Sperm whale myoglobin (1MBD).

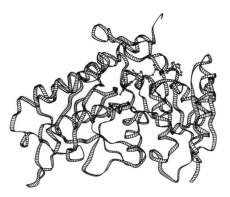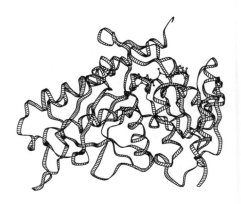

Figure A92. Malate dehydogenase (4MDH).

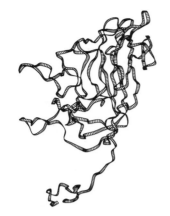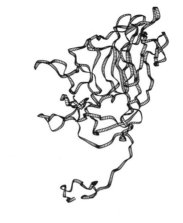

Figure A93. Mengo virus VP1 (2MEV).

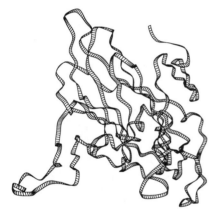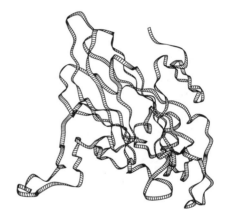

Figure A94. Mengo virus VP2 (2MEV).

Figure A95. Mengo virus VP3 (2MEV).

Figure A96. Mellitin (1MLT).

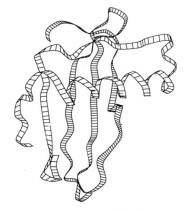
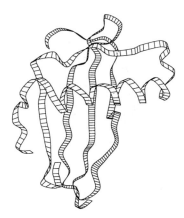

Figure A97. Monellin (1MON).

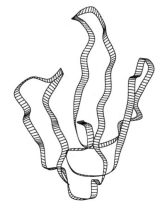
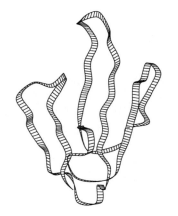

Figure A98. Neurotoxin B (1NXB).

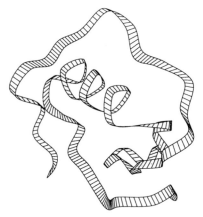 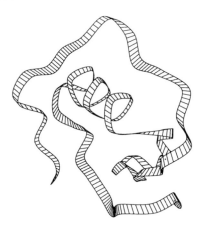

Figure A99. Ovomucoid third domain (2OVO).

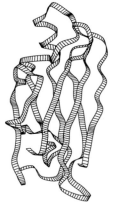 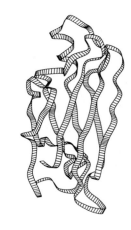

Figure A100. Prealbumin (2PAB).

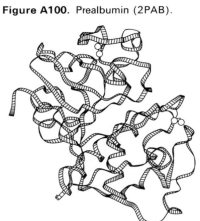 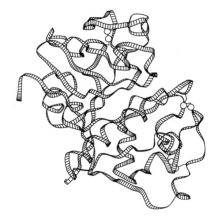

Figure A101. Papain (9PAP).

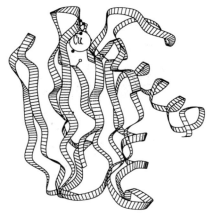
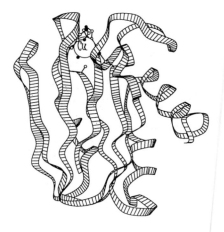

Figure A102. Pseudoazurin (2PAZ).

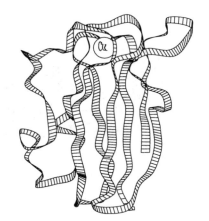
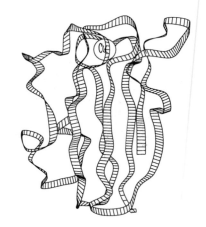

Figure A103. Plastocyanin (1PCY).

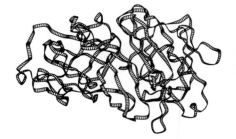

Figure A104. Pepsin (4PEP).

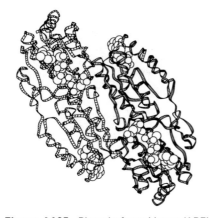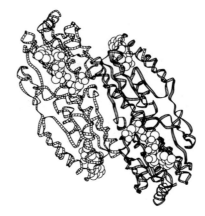

Figure A105. Phosphofructokinase (1PFK).

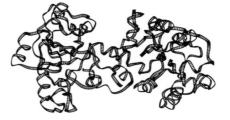

Figure A106. Phosphoglycerate kinase (3PGK).

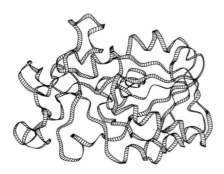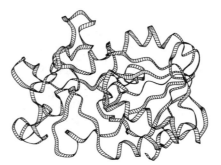

Figure A107. Phosphoglycerate mutase (3PGM).

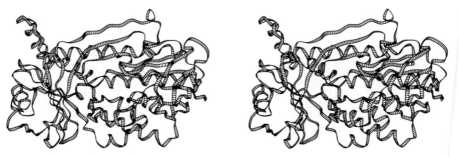

Figure A108. *p*-hydroxybenzoate hydroxylase (1PHH).

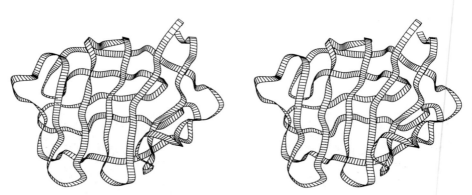

Figure A109. Photoreactive yellow protein (1PHY).

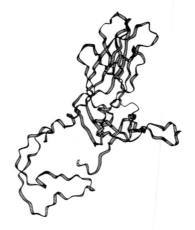

Figure A110. Polio virus VP1 (2PLV).

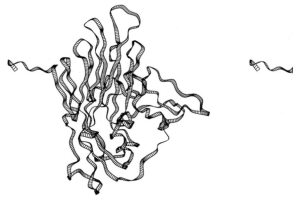

Figure A111. Polio virus VP2 (2PLV).

Figure A112. Polio virus VP3 (2PLV).

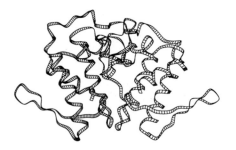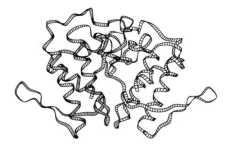

Figure A113. Snake venom phospholipase (1PP2).

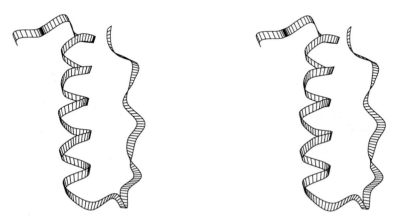

Figure A114. Avian pancreatic polypeptide (1PPT).

Figure A115. Photosynthetic reaction center *R. viridis* (1PRC). (a) C subunit (cytochrome). (b) L subunit. (c) M subunit. (d) H subunit.

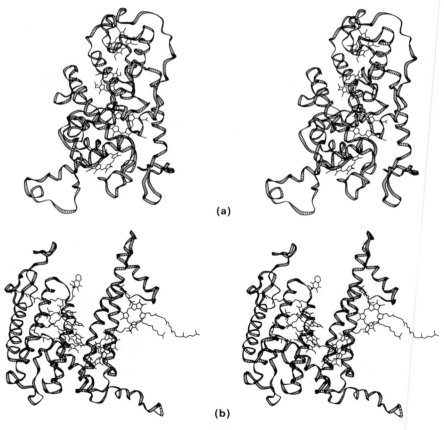

(a)

(b)

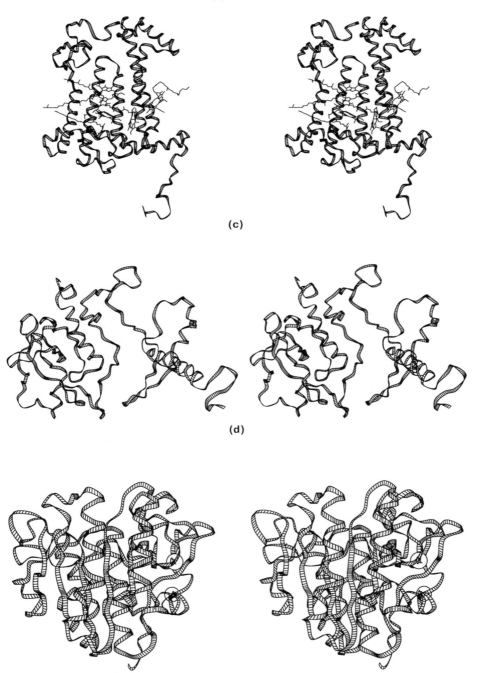

(c)

(d)

Figure A116. Proteinase K (2PRK).

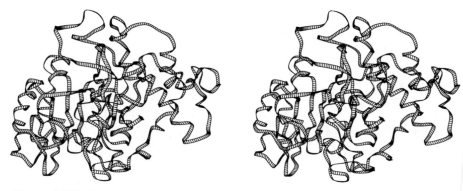

Figure A117. Carboxypeptidase/transpeptidase (1PTE).

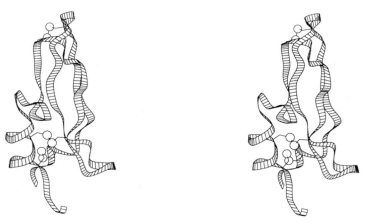

Figure A118. Bovine pancreatic trypsin inhibitor (5PTI).

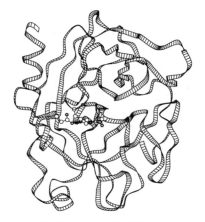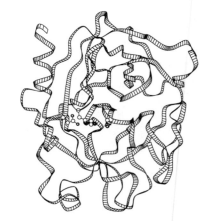

Figure A119. Trypsin (4PTP).

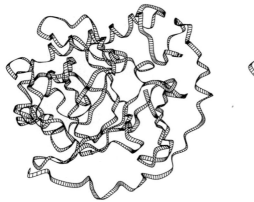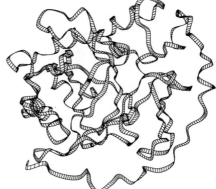

Figure A120. Pyrophosphatase (1PYP).

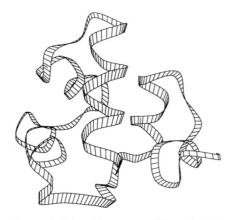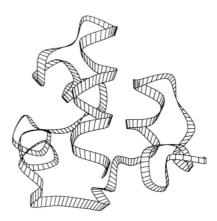

Figure A121. 434 repressor (N-terminal domain) (1R69).

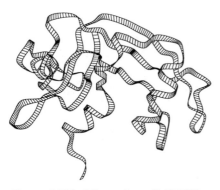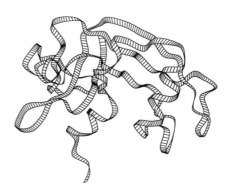

Figure A122. Ribonuclease B (1RBB).

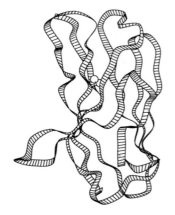
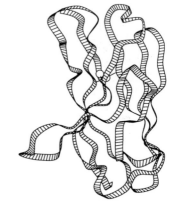

Figure A123. Immunoglobulin V$_\kappa$ domain (1REI).

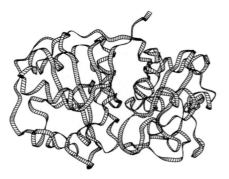
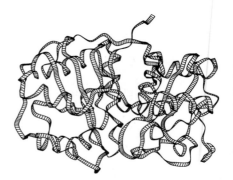

Figure A124. Rhodanese (1RHD).

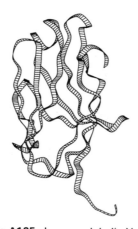
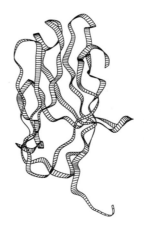

Figure A125. Immunoglobulin V$_\lambda$ domain (2RHE).

Figure A126. Rhinovirus VP1 (4RHV).

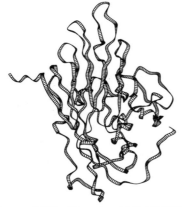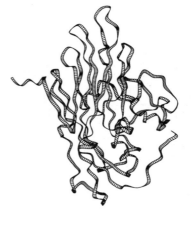

Figure A127. Rhinovirus VP2 (4RHV).

Figure A128. Rhinovirus VP3 (4RHV).

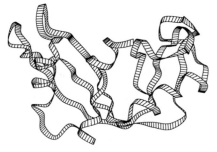
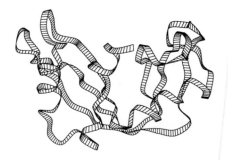

Figure A129. Ribonuclease A (1RN3).

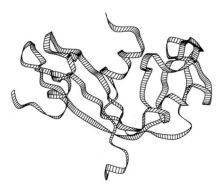
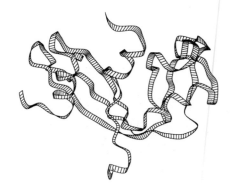

Figure A130. Ribonuclease S (1RNS).

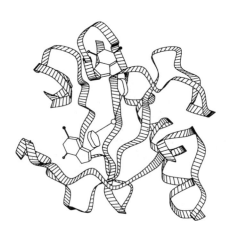
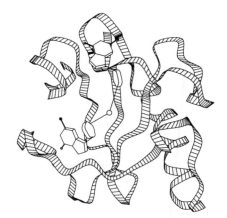

Figure A131. Ribonuclease T_1 (2RNT).

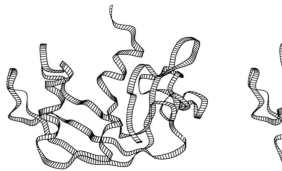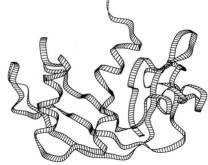

Figure A132. Rat mast cell protease (3RP2).

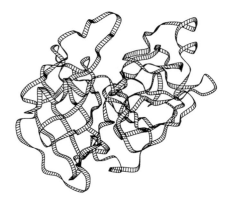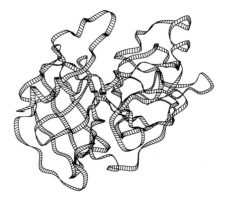

Figure A133. Ribonuclease A (7RSA).

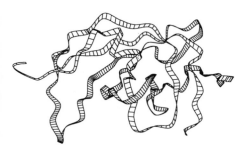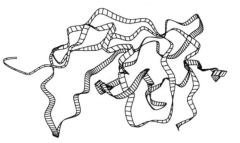

Figure A134. Rous sarcoma virus protease (2RSP).

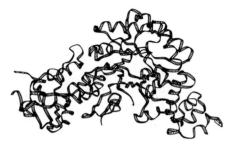
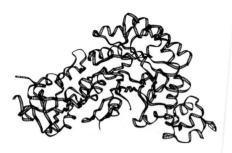

Figure A135. Rubisco (2RUB).

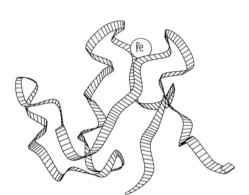
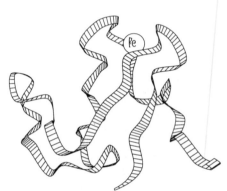

Figure A136. Rubredoxin (5RXN).

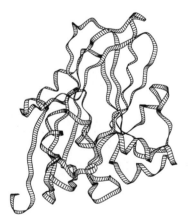
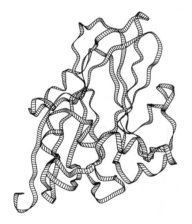

Figure A137. Southern bean mosaic virus (4SBV).

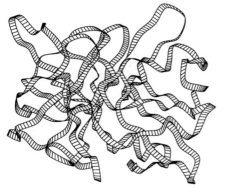
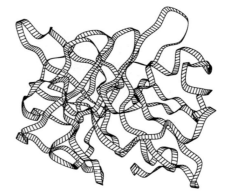

Figure A138. *S. griseus* proteinase A (2SGA).

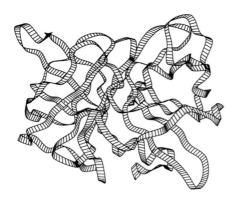
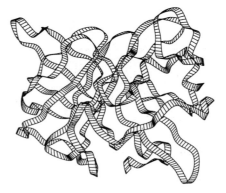

Figure A139. *S. griseus* proteinase B (2SGB).

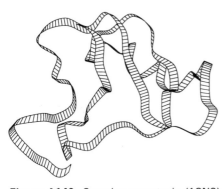
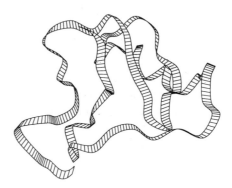

Figure A140. Scorpion neurotoxin (1SN3).

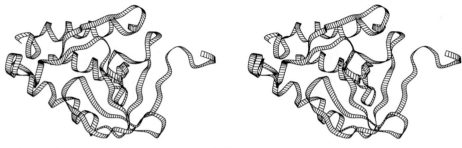

Figure A141. Staphylococcal nuclease (2SNS).

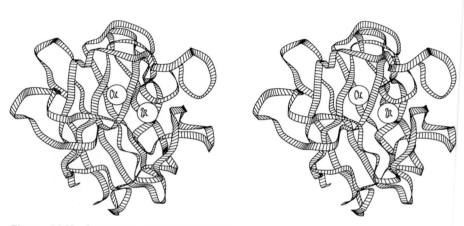

Figure A142. Superoxide dismutase (2SOD).

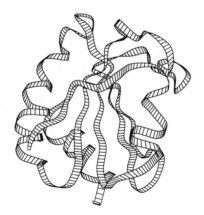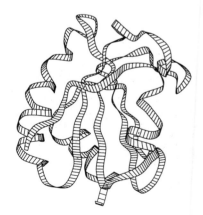

Figure A143. Thioredoxin (1SRX).

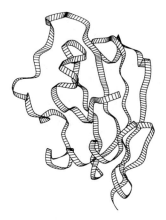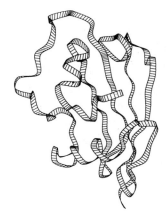

Figure A144. Streptomyces subtilisin inhibitor (2SSI).

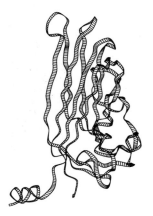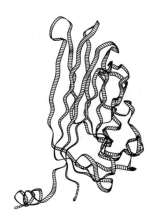

Figure A145. Satellite tobacco necrosis virus (2STV).

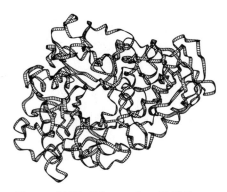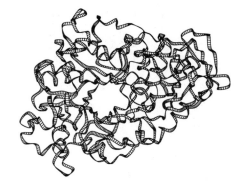

Figure A146. Taka-amylase (2TAA).

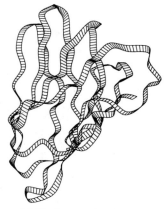 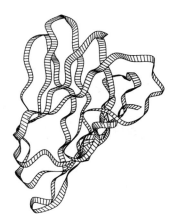

Figure A147. Tomato bushy stunt virus (2TBV).

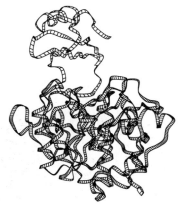 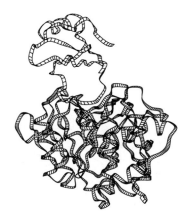

Figure A148. Thermitase/Eglin-C (1TEC).

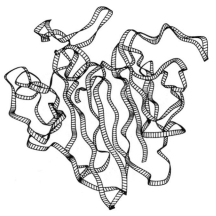 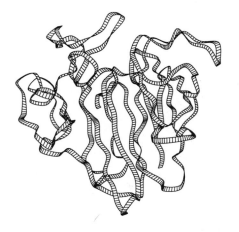

Figure A149. Thaumatin I (1THI).

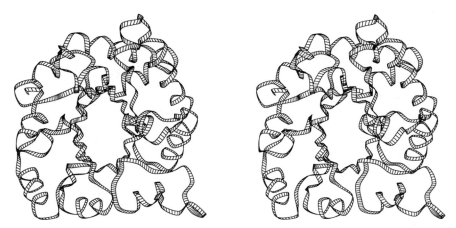

Figure A150. Chicken triosephosphate isomerase (1TIM).

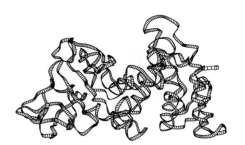

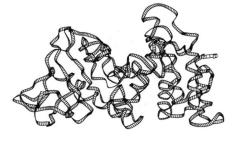

Figure A151. Thermolysin (3TLN).

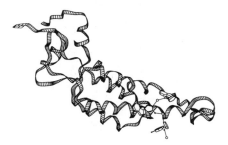

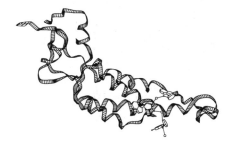

Figure A152. Tobacco mosaic virus (2TMV).

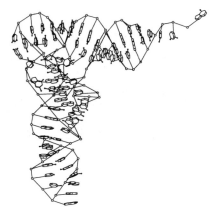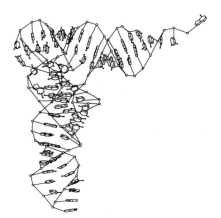

Figure A153. Yeast tRNA^{Phe} (4TNA).

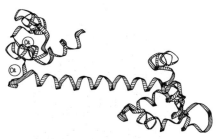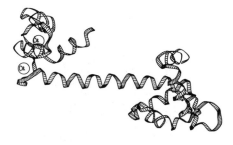

Figure A154. Troponin C (4TNC).

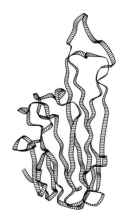

Figure A155. Tumour necrosis factor (1TNF).

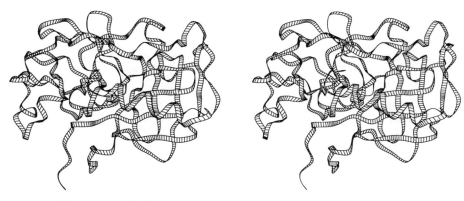

Figure A156. Tonin (1TON).

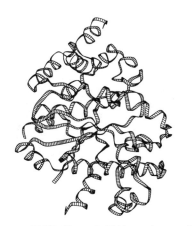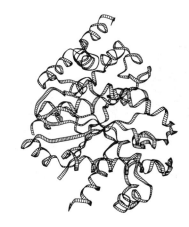

Figure A157. Tyrosyl tRNA synthetase (2TS1).

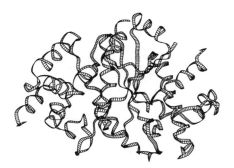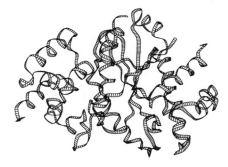

Figure A158. Tyrosyl tRNA synthetase (4TS1).

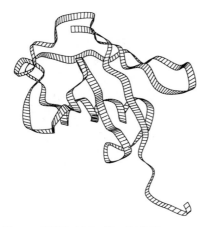
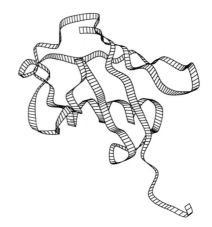

Figure A159. Ubiquitin (1UBQ).

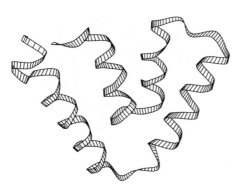
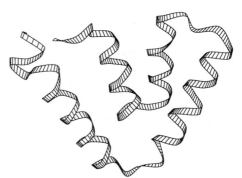

Figure A160. Rabbit uteroglobin (1UTG).

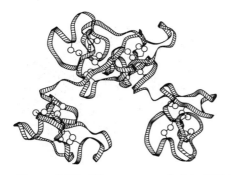
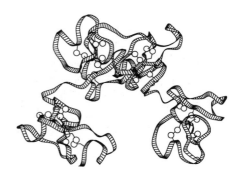

Figure A161. Wheat germ agglutinin (3WGA).

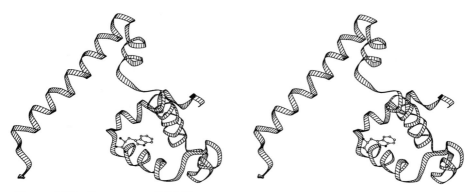

Figure A162. Trp repressor (1WRP).

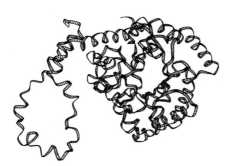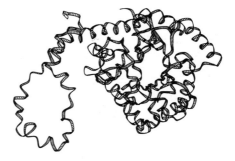

Figure A163. Xylose isomerase (4XIA).

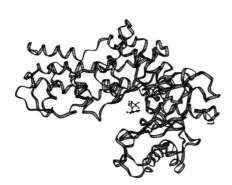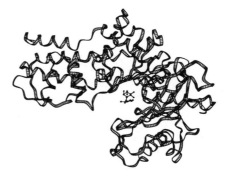

Figure A164. Hexokinase (2YHX).

Index

Bold-face numbers indicate pages containing pictures.